We Live in the Water

WE LIVE IN THE WATER

CLIMATE, AGING, AND SOCIOECOLOGY ON SMITH ISLAND

Jana Kopelent Rehak

JOHNS HOPKINS UNIVERSITY PRESS
Baltimore

© 2024 Johns Hopkins University Press
Images except on pages xviii and 78 © 2024 Jana Kopelent Rehak
All rights reserved. Published 2024
Printed in the United States of America on acid-free paper
2 4 6 8 9 7 5 3 1

Johns Hopkins University Press
2715 North Charles Street
Baltimore, Maryland 21218
www.press.jhu.edu

Library of Congress Cataloging-in-Publication Data

Names: Rehak, Jana Kopelentova, 1968– author.
Title: We live in the water : climate, aging, and socioecology on Smith Island / Jana Kopelent Rehak.
Description: Baltimore : Johns Hopkins University Press, 2024. | Includes bibliographical references and index.
Identifiers: LCCN 2023028585 | ISBN 9781421448428 (paperback) | ISBN 9781421448435 (ebook)
Subjects: LCSH: Smith Island (Md. and Va.)—Social life and customs. | Smith Island (Md. and Va.)—Environmental conditions. | Human beings—Effect of environment on—Smith Island (Md. and Va.) | Human ecology—Smith Island (Md. and Va.) | Climatic changes—Social aspects—Smith Island (Md. and Va.)
Classification: LCC F187.C5 R44 2024 | DDC 304.209752/23—dc23/eng/20230802
LC record available at https://lccn.loc.gov/2023028585

A catalog record for this book is available from the British Library.

The image on page xviii is used courtesy of Jennings Evans.
The image on page 78 is used courtesy of Mitzi Brimer.

Special discounts are available for bulk purchases of this book. For more information, please contact Special Sales at specialsales@jh.edu.

Contents

Preface *vii*

INTRODUCTION The Politics and Poetics of the Weather World   1

1 Weather Is Everything   34
2 Ways of Knowing   57
3 Land and Water   78
4 Shifting Grounds   96
5 Broken Bodies   117
6 The Taste of Things and Comic Relief   140
7 The Art of Creative Futures   165

EPILOGUE Ethnographic Poetics   190

Acknowledgments   201
Bibliography   203
Index   213

Preface

GRASSLAND

Travelers to Smith Island take a boat, leaving behind Maryland's Eastern Shore and heading toward the beginning of the marshlands in the lower Chesapeake Bay. Once in the marshlands, it is necessary to follow the waterway that is marked out by wooden poles along the seascape path to the island. The route passes strips of grassland, animated by the winds, that change colors from green to yellow and then back to green again each day and season. The grass, like water in the ocean, is always in motion. The boat pushes against the wind, and as it moves closer to Smith Island, the seascape's salty air begins to mix with the sweet smell of marshland. As it turns into the channel, the boat passes the first red-painted shanty sitting close to the harbor. It is embellished with an American flag and fronted by crab pots piled up on the wooden deck. Farther in are the workboats docked by the pier, which delineates a harbor. This is Ewell, a village built along the northeastern coastline of Smith Island.

On the island, a short road leads to a marina and then to what is the busiest crossroad in Ewell. The Methodist church, with its white tower making it the tallest building in town, is surrounded by a graveyard, and the preacher's house is next to it. There is a marked contrast between the church and the modern, light-colored-wood architecture of Smith Island's museum. Beyond is a red-painted building—the old Ruke's Restaurant, now closed—sitting out over the marshland. Older two-story houses, painted white, some of which are occupied and some empty, stand next to newer one-story homes, where most of the current Smith Islanders live.

A road connects this part of Ewell to the South End, known on the island as *down by the field*, where there is a small post office and a bakery tucked between the houses. At the North End, known as *up on the hill*, there is a cluster of houses and an electric plant. Another road leads straight through the village, passing a school with a playground, a few scattered houses, a firehouse, and a social hall. It then continues, winding out into the marshland, and connects Ewell with Rhodes Point, often flooded recently by water at high tide. Rhodes Point is a one-street village where houses, the church and graveyard, a dental clinic, and a senior center face the waterfront and are exposed to the wind and rising waters from the western shores of the Bay.

From Rhodes Point or Ewell one can continue by boat through narrow marshland channels to Tylerton, a little village on another island in the southern part of the Smith Island archipelago. Tylerton—which features a cluster of houses surrounding the church and graveyard, a shop with a coffeehouse, and a tiny post office—is right on the waterfront. In Tylerton, the view of the southern edge of the village, like that of the other two communities, is softened by the marshlands and perfumed by an earthy scent carried across the island when the wind is blowing from the south. Walking along this path, I always notice how suddenly my body drops into island time, what new island residents think of as the tension in one's shoulders leaving and one's senses surrendering to the sounds carried by the wind, the splash of boats rocked by the water, and the cries of seagulls.

Smith Island is the home of the largest island community in Maryland's Chesapeake Bay. Today, the island's fewer than 200 people are divided among three distinct towns: Ewell, Tylerton, and Rhodes Point. According to the kinship records in the Jennings Evans archive in Maryland's Somerset County Library, Smith Island descendants trace their history back to the original English, Welsh, Cornish, Scottish, and Irish settlers, who arrived in the late 1600s, but also to the subsequent Dutch and Spanish ones. After these early settlers moved to Smith Island from Virginia and Maryland's Eastern Shore, they traditionally sustained themselves on the island by small-scale farming, gardening, fishing, and hunting, and later by commercial crabbing and oystering. Islanders, and others, refer to their occupation as being *watermen*, a term widely used in the Mid-Atlantic area to denote those who derive their livelihood from working on the waters of Chesapeake Bay, the largest estuary in the United States. Smith Island's three communities share kinship and historical patterns, are differentiated somewhat by socio-geography, but collectively belong to the Bay's socioeconomic system, which relies on commercial crabbing and oystering.

Historically, watermen used sailboats, called *skipjacks*, and left home for days to sail north in Chesapeake Bay, searching for blue crabs during the summer season (between April and October) and oysters during the winter (between November and March). Teenage males joined their kin to labor on sailboats in the Bay for several weeks at a time. The role of women was structured to support their husbands, sons, or brothers by managing affairs in the community and family life. Thus, from an early age, boys were trained by their kinsmen to work on the water, while girls were taught to take care of collective life on the island. Today, women's work on the island remains critical, and men often emphasize that they can't manage without the women.

In the islanders' narratives about the past, they tell stories of extraordinary times of hardship during America's Revolutionary War, when the British army used the island for its food resources. Many of the stories, however, mention the prosperity of the community's craftsmen, such as blacksmiths, dressmakers, and barbers, due to the geographic position of Smith Island at the crossroads of ships' channels to upper Chesapeake Bay. The development of a railway on Maryland's Eastern Shore at the end of the nineteenth century, which was a major change in the Chesapeake Bay area, meant that Smith Island was no longer the hub of major commercial routes. But the high demand for Bay oysters created new economic prosperity on Smith Island, reflected in population growth records and the construction of larger houses dating from this time. Some of the older islanders still remember what it was like growing up in cold and drafty old island houses, many of which have now been abandoned and left to slowly decay as people moved away from the island or built smaller one-story homes during the twentieth century.

Several times in the past, the members of some families were forced to leave the Smith Island community when the crabbing and oystering economy was in decline. "They were hard-hit during World War I and World War II," Jennings, a Smith Islander, explained when he spoke to me about local history. His family had moved from the island to Baltimore in order to economically sustain themselves during World War II. Working on the water in the Bay changed drastically after this war, when many watermen transitioned from skipjacks to working on motorized boats. It gradually changed again during the 1980s, when new environmental regulations for watermen were put in place. Even though Smith Island is quite remote from the mainland, it has never been isolated. From its founding, the community existed at the crosswords of maritime trade, was part of the history of pirates' existence in the Bay, lay on the path of Methodist missionaries, and was later exposed to global processes. *Island ways* (the term islanders use to identify their practices) have continued, yet always under conditions of flux. Therefore, long-standing island ways are always subject to change, which is reflected in the oral history, local journaling, newspapers articles, and books that are based on a collaboration between members of this community and outsiders.

Contemporary residents are concerned about ecological changes related to flooding and land erosion. Smith Island's ecology, like that of many other small coastal islands, has been affected by a slowly shifting climate. As sea temperatures have risen, the crabs and oysters that traditionally sustained this community have started migrating north. Some islanders openly claim disbelief in climate change, while others observe and address their problems with water, land, and health as part of the aging process. Because traditional work on the water—crabbing and oystering—within this community has been diminishing over time, many Smith Islanders have moved to the mainland. Nonetheless, despite the absence of younger watermen and the heavy burden of current work regulations, the mostly older population of Smith Islanders is finding that crabs have increased in value and thus are selling them to high-end consumer markets in Washington, DC, and New York. The community is also developing a more robust tourist economy in tandem with their traditional fishing economy.

In addition to the original families, which form the core of this community, in the past ten years new residents with diverse backgrounds—retired couples, a few single individuals, and families—have moved to the island from mainland. Some newcomers belong to the community by marriage, while others come to the island in search of a better place to live or as retirees. Some of these new residents remain on the island full time, while others only use their island houses seasonally or for short vacations. People who moved to the island as adults do not engage in traditional fishing jobs, but many participate in the tourist economy through their knowledge of local traditions, as well as provide services to the island community. In contrast to island-born residents, who are focused on keeping or reinventing their traditional life, these newcomers search for a way of life that differs from what they had on the mainland. What they desire on Smith Island is a "peaceful life" (as some put it) and acceptance from the island community. It is significant that on their way to becoming an integral part of the community, these newcomers contribute to the island's future.

This book is about people who endure in what is referred to as their *weather world*, and it emphasizes how people reinvent their lives in light of socioecological change. I do not seek to answer questions about the politics of climate change, nor is it my aim to stay within the limits of literature about fisheries. Instead, I try to make sense of Smith Islanders' own aging experiences, perceptions, and knowledge—that is, their ecological, cosmological, and sensory knowledge. This book is an invitation to view the human dimensions of the island's climate and ecology. I explore what it means to live in the weather world on the island, which is affected both by the slow process of climate change and rapid socioecological changes. My narratives offer new insights into how the people known as Smith Islanders view their own aging and their ever-changing environment. In examining how they engage during their life course with their environment—that is, their land and the surrounding seascape—I consider their ecological knowledge in relation to their work and place, their collective life, and the ways in which they see their subjective experiences. My hope is that readers will learn about Smith Island's men and women, as well as their ways of being in their weather world as they are growing older.

In the anthropology classes I teach, students often ask why people like Smith Islanders continue living in geographically challenging areas that are harshly impacted by volatile weather conditions and the longer-term effects of climate change. Each chapter in this book provides ethnographically grounded discussions showing how people construct their sense of self on a small island through their engagement with their environment and the community. My aim is to show how people like Smith Islanders manage their lives in an environmentally unstable place. I investigate how, through the process known as *placemaking*, they are reinventing their local traditions and practices by engaging with their environment and their weather world. I argue that changing environmental conditions and vanishing social traditions lead us to see the sadness arising from loss, yet I emphasize how Smith Island people continue their life and maintain future hopes for their community, regardless of environmental and aging hardships.

Smith Islanders' socioecological realities raise a number of questions about political agency, sustained by poetics embodied in people's sense of self and place. In any context, the idea of being forced to relocate from one's inherited island community is abhorrent. Smith Island is a clear example of a community being transformed. Some traditions are being forced into extinction, yet many islanders still hold on to the hope that, by reinventing old traditions, their water-based lives will continue well into the future. Islanders descended from the original settlers, together with newcomers, are developing new practices related to work and social rituals, as well as benefiting from the public education manifested in tourism. While this current generation is redefining what it means to grow old on the island, they are overwhelmed by both the loss of their kin (through death or relocation to the mainland) and the loss of their land (to erosion caused by frequent flooding). Yet they trust their confidence, shared by others in this community, in their ability to manage their way of life, since they believe their health and prosperity stem from the fact that they are ordained by God to continue forward, maintaining their resilient nature and their manner of survival. Therefore they move forward into the future with a Christian faith in God, constituted by their knowledge of self and place.

Smith Islanders are poetic and creative people; they find poetics in the seascape and in the storytelling embodied in ordinary talk as they manage economic realities. Language play is often part of their daily conversations (Limón 1994). Smith Islanders have also used diverse written traditions to capture different types of knowledge. They attend to the daily task of maintaining logbook records about weather patterns, the behavior of crabs, and variations in oyster seasons, but some islanders also keep journals, reflecting island life in their short stories and poetry. Collective scriptwriting is part of their preparation for traditional performances at annual celebrations. Family photographs and texts, received as gifts or exchanged by islanders, are arranged in albums or displayed on the

HARBOR

walls of island homes. These are pictorial kinship trees that gain new meanings at a time when younger kin are leaving the island and older islanders are dying. Historically, creativity was part of their regular work. Men have been skilled in crafts such as building a house or shanty building, carving decoys, and making boat models, and women in cooking, baking, and various handcrafts, such as knitting and quilting. But perhaps the ultimate poetic and creative aspect of life on Smith Island, the thing that most firmly binds Smith Islanders together, is their collective oral traditions, like storytelling, joking, and singing during social events, both in church and at annual celebrations.

"The nature and quality of what anthropologists learn is profoundly affected by the unique shape of their fieldwork," as Lila Abu-Lughod (2016) reminds us in her ethnography, *Veiled Sentiments*. Anthropologists seldom write about their specific sensory, emotional, spiritual, or other personal experiences during fieldwork. Yet creating a new space for fieldwork poetics and expressions of feelings has become increasingly acceptable in ethnographic narratives (e.g., Jackson and Piette 2015; Lambek 1981). I relate to the Smith Islanders through the creativity I bring with me as an artist and anthropologist, but also by sensory experiences I have when I stay on the island, as well as through ordinary conversations and small talk about my family or the well-being of others.

My fieldwork on Smith Island is always interrupted by the trips back to my home in Baltimore, yet I noticed that after some years, I developed new ways in my life, adapting to the island's habits. My experiences there changed how I engage with the weather world and relate to people in general. I watch the weather in different ways. I observe the wind and bring sensory memories from the marshland into the city. I also long for a sense of community. While traveling back to the city, I hold on to the energy I gained from a sense of belonging on the island. I value friendships with many of the islanders, and I call them when I fear my own weather worlds. Seeing how Smith Islanders pursue their lives—with focus and yet with playful poetics of storytelling and joking, many times simply noticing the sounds of birds, the sweet smell of honeysuckle or pungent smells from marshland, the behavior of crabs and other animals in the seascape, while still checking on the pain of their fellow islanders—enriched my own life. After I lived through summer storms, humidity, heat, strong winds for days, flooded roads, and attacks by green flies and mosquitos with every step outside, I began to wonder if hardships become insignificant after one becomes accustomed to a particular environment.

I saw how tired watermen are when they return home from the sea day after day. I marveled at women's strengths when they extended their workdays into the night or waited for the wind to die down. I began to realize that people who work on the water all their lives are humbled by the limitations of the weather, as well as challenged by the decreased vitality of their bodies. Life on Smith Island teaches people every day about their own limits, as well as the unpredictable and precarious realities of human existence. This book is my ethnographic story about the Smith Island people and their changing weather world, circumscribed by the years between 2013 and 2021. As I tell their stories, I present the islanders' collective and individual ways of being and knowing, composed of experiences, beliefs, and emotions that are often hidden in the shadows of the worldwide political discourse involving global climate change.

As I continue my visits to Smith Island, I am aware that my writing captures the social life on the island during the specific period of my fieldwork there (2013–2021). A lot has changed on the island since 2021, when I stopped my systematic data collecting about island life. Many older islanders that I knew well have either died or now live in a nursing home, and a number of newcomers whom I don't know have moved to the island. Aggressive flooding is causing new challenges for both long-time residents and new island people. Water levels are rising more often, and flooding or high tides not only are damaging houses and land on a greater scale, but also prevent islanders from freely moving around the island. When water surrounds one's home, its inhabitants are held hostage and become prisoners. I wrote this book with compassion for the people of Smith Island, with respect for their knowledge about their place and their weather world and for the hope they carry for their island's future, despite its uncertain grounds.

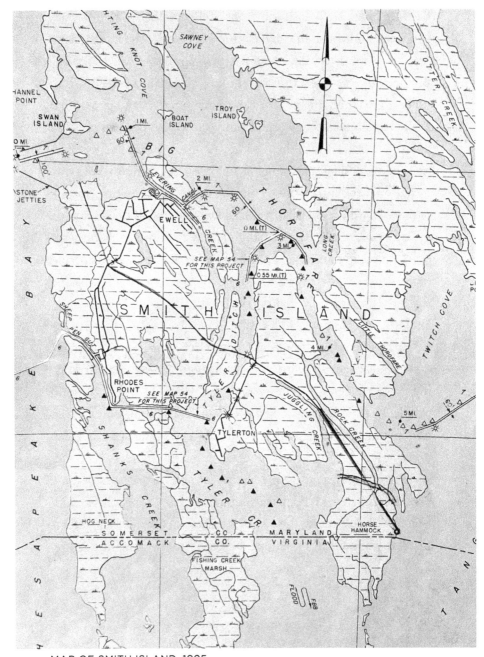

MAP OF SMITH ISLAND, 1985

We Live in the Water

Introduction

THE POLITICS AND POETICS OF THE WEATHER WORLD

"They care more about birds than the lives of the people in the Bay," one of the Smith Island women told me in a conversation about the turmoil the islanders experienced after Hurricane Sandy in 2012. When I arrived on the island in 2013, I found that the topic of relocation from the island to the mainland was often raised during my first conversations with the residents. Governmental oversight adversely impacted Smith Islanders' lives after Hurricane Sandy. Official reactions to the devastation of the island's vulnerable shoreline by this natural disaster were politicized and caused considerable communal distress. Smith Islanders' family albums hold a plethora of photographs documenting the destruction on the island from Hurricane Hazel in 1954. These snapshots show the broken houses and speak to a history of volatile weather conditions. But, unlike Hazel, Hurricane Sandy provoked questions about the island's future. Smith Islanders were expected to quickly establish a plan of action, which would ultimately be approved by outsiders. Faced with possible relocation to the mainland, and in the absence of local government, the islanders formed a new civic organization, Smith Island

United. It would allow the community to creatively address the island's future sustainability and resilience. The subject of climate change is not comfortable for Island residents, but it flared into prominence in 2012, after Hurricane Sandy. Out of a desire for resiliency, and to continue the life they inherited, the islanders developed a strategic plan for the development of a tourist economy, in order to diversify their long-established traditional income from crabbing and oystering. While securing funding for land improvement is the priority and focus of Smith Island United, the organization also exercises oversight of diverse local issues related to health care and other community services, as well as to fishing and the tourist economy.

When I asked residents about the events stemming from the hurricane's aftermath, I noticed frustration growing out of the locals' feelings that their preferences and concerns were misunderstood. "People were ready to take care of repairs themselves, but the community was short of resources to repair the damage," said one of the islanders. Different residents shared their concerns, and I could sense anxiety, driven by the threat of losing the island's land, in these narratives. For example, John noted:

> During the aftermath of Hurricane Sandy, the government offered everyone who lived here the option of being bought out on the condition that once the property was sold, it would be marked forevermore a "dead property." That meant no one could ever live there again, and the houses would be demolished. This was heartbreaking to many. They were basically trying to phase out Smith Island for good, to sound its death knell. It caused a media backlash and lots of attention for the island. People came to our defense and now we have this government agency in charge of deciding how a ton of money given to us will be used.

During multiple conversations I could feel the distress of the community. They felt betrayed, since the government's response to damage from the disaster was to ask all residents to relinquish their way of life on Smith Island. It was a painful surprise, as many indicated. All three of the local communities shared a sense of injustice at possibly losing their right to remain on their island. One resident expressed his view this way: "When time comes for the USA government to support hazard-prone communities from rising sea levels, Smith Islanders should be part of such a project." To him, governmental actions with regard to geographic areas need to be undertaken with a strong protection of equal rights for *all* US citizens. In conversations about climate change and Smith Island's future, people emphasized, "Our main problem is erosion and that is our focus."

"Politics is in everything and politics rule," Elmer (known as "Junior") said. Governmental support for building resilience in coastal communities is now conditioned by the politics of climate change. Confronted by such policies, Smith Islanders, like many others in coastal communities, claim the final authority over their land. A united voice proved to be a powerful expression of collective agency when the island's habitat was threatened by the crisis that followed the hurricane. It became evident to me that islanders were outraged by a proposal, part of a long-range plan, to make Smith Island a wildlife refuge by removing people from the island. In our conversations, some *watermen*—a term widely used in the Mid-Atlantic area to denote those who derive their livelihood from working on the waters of Chesapeake Bay—expressed their frustration over what they see as the long-term politics of assigning environmental blame. "There is a negative view of watermen outside," Buddy, one of the younger watermen, told me. "We are seen as bad for the Bay, and yet I don't know why. It doesn't make sense. We depend on our environment for survival. To harm life in the Bay is against our logic, we depend on the Bay," he concluded. Fishing regulations, according to watermen, ignore their hard-earned knowledge. "Oystering is like gardening, you have to care by returning back when growing oysters," explained Junior, Buddy's father. When I asked him about oyster farming, Junior said, "That is difficult, very small chance you will succeed economically. It is lot of work and not much chance and they pay less. Market demands wild oysters over farm raised and they pay better."

From their voices, I could hear how a sense of frustration is added to their uncertainty, conditioned by the weather world. It is evident from my conversations with them that restrictions from outside authorities, who are in some instances viewed by the islanders as ignorant of the residents' expertise, evolved over time into politically driven tensions. Given the disconnect between local people and state authorities, the issue is more about the knowledge/power/authority struggle than it is about environmental knowledge itself. Smith Islanders, like others who live in small rural communities, are increasingly more dependent on federal and state agencies, as many of their resources come from outside the community, yet they also realize the importance of their collective agency. The politics of climate change, initiated by outsiders, are perceived by Smith Islanders as a threat. To them, the island holds biographical histories that are closely intertwined with the evolving heritage through multiple modalities. It is in the poetics and knowledge of their everyday lives that they bond with their island. Living on the island is, for many, "the art of the life," as Chris put it. It would be impossible to replace that art after a move to the mainland.

Life on the island teaches people about the unpredictable and precarious realities of everyday human existence. Armed with their accumulated local knowledge and their faith in God, they are suspicious of generic models of management. Smith Islanders have experienced governmental oversight regarding hunting and fishing practices, and this has struck deeply against their sense of survival. Federal and state regulations for fisheries play a significant role in watermen's lives in the Chesapeake Bay area. "We are regulated already by the weather. In the winter you may not go out on the water for weeks, because of dangerous weather. During the summer season the heat prevents you from working for long hours. Being an experienced waterman also means that you know your physical limits. You can't risk your life and push yourself over the limits; every waterman knows his limits," said one of the watermen in a conversation about these regulations. The ever-present uncertainty Smith Islanders endure is rooted in their relationship with death (Allison 2013). Death is not a strange phenomenon in an islander's life.

As Jennings told me, consideration of locally specific ecological knowledge and coastal peoples' abilities to "move in different pace" will not suffice. Because Smith Islanders embrace such uncertainty in their day-to-day weather world, their future planning often evolves through their assessment of ecological changes and their economic readjustments. Their imagined futures are always constituted in present practices. In efforts to sustain their short-term survival, life on the water involves more than just economic success. "You must be observant about what your father knew and watch the patterns of crab movement. Go hunting for what we catch, being your own boss, and going when and where you please is almost as important as the money," said Chris. For the islanders, dismissing their views, which are informed by their ecological, social, and cosmological knowledge, is to dismiss their sense of self, which has been realized through their way of being in the world. Because Smith Islanders embrace the uncertain in their day-to-day weather world, their future planning often evolves through their dynamic assessment of ecological changes and their economic readjustments.

From my conversations with Smith Islanders, I encountered the tensions between the logic of their ecological knowledge and the political discourse of climate change. I could see how, for them, their purpose in life is connected to their sense of inherited place, and how their heritage gives them purpose and a focus to their lives, regardless of any hardships they must endure. It is from accounts like these that I challenge current popular theories about the impact of climate change on a small community. Chesapeake Bay watermen understand sustainable biological and social processes very well, and that is why they so often resist attempts by state authorities to impose regulations (Paolisso 2003; Paolisso et al. 2006). The politics of climate change show a clear disconnect between people in small coastal communities whose cumulative local knowledge has been learned over time and scientific experts from outside who claim to be authorities on environmental knowledge (Berkes 2015; Ponkrat and Stocker 2011; Roscoe 2014).

Examining the socioecological realities of people living on Smith Island raises many broad questions about the places humans inhabit and their climatic futures. In this weather-dependent island community, sustainability has always been precarious, and people there have lived in very uncertain conditions. They assert their agency in *placemaking*—to them, a gift, in that it creates a sense of belonging and nourishment. Emotional experiences, manifested in anxieties arising from pain, health problems, weather, economics, or death, are creatively expressed in social contexts, where they are transformed by collective prayers, storytelling, or dramatic performances. When Smith Islanders speak about purpose in their lives and reasons to hold on to hope for their community's future, they emphasize their joy from the simple things in day-to-day existence and their trust in God to provide them with all they need in order to survive on the island. In conversations about climate change, Smith Islanders emphasized, "If God wants us here, He will provide, and if He doesn't, He will show us our way and what to do."

It is from this position of faith that the islanders acquire peace and confidence in their own adaptation to climatic and other changes.

Introducing Smith Islanders' beliefs and emotions, derived from their experiences, will not settle the political discourse begun after Hurricane Sandy. But as these emotions and views continue to underlie the conditions in this community, it is helpful to contextualize the narratives presented in this book. Historically, scientific interests in climate mostly addressed what occurred in the past, as David Harvey and Jim Perry (2015) recognize, along with others concerned with climate and heritage studies. The current climate change discourse, however, is oriented toward climatic futures, which are easy to imagine for lowland island communities. Yet for Smith Island people, I argue that the topic of hope is very important, because it helps them redefine and reinvent their practices in relation to their environment. Many are in the later stages of life, yet they continue to cope with their weather world as best they can, regardless of limits posed by severe flooding and their own diminishing vitality. As they reinvent their traditional way of life, which they regard as their heritage, they also welcome modern technology and newcomers, as both are engaged in reconstructing the island's future.

Socioecology and Phenomenological Anthropology

Throughout the world, the effect of weather on daily life is one of the most shared human experiences. In the short term, one's way of life is always situated in a particular weather world. Yet the long-term effects of global climate change on local communities is also a source of stress on socioecological systems, influencing agriculture, fishing, forestry, and water. In small communities affected by changing socioecology, an awareness of people who have died and land

HARBOR IN EWELL

that is no longer what it was are increasingly part of their existence. As I observe the social life of people who live on Smith Island, I describe genuine moments in their subjective experiences, as these are part of their existing conditions, which are integrated into larger ecological systems. Showing how Smith Islanders develop their worldviews, I provide insight into how they relate to their place and why moving from the island is not considered to be an option *for many islanders* even when environmental circumstances have increased their vulnerability. How *people* maintain their way of life despite near-impossible conditions is an important question many anthropologists ask in the context of a sudden crisis in people's lives, such as one triggered by a natural disaster or war. But the slow degradation of one's environment and social life requires a long-term focus and strategies reinventing what is the cause of despair.

The current, fast-growing spate of anthropological studies on the impact of climate on humans have focused on local populations in small rural communities. These, in turn, have provided evidence for the need to address the sustainability of such places, which have been affected by global environmental changes. Ecological and environmental anthropologists addressing the alarming shifts in diverse socioecological systems worldwide agree that coastal ecosystems are some of the sociobiological systems most affected by global climate change (Anderson 1996; Crate and Nuttall 2016; Marino 2015; Ponkrat and Stocker 2011; Roscoe 2014; Strauss and Orlove 2003). Collectively, many of these studies have used ecological unity as a platform, and they generally emphasize the importance of local traditional knowledge when addressing an environmental crisis in coastal areas worldwide. Socioecological research suggests that it is critical to reconnect social (i.e., human) and ecological (i.e., biophysical) systems (Berkes 2015; Fiske 2016; Ingold 1986; Paolisso et al. 2006; Ponkrat and Stocker 2011; Roscoe 2014).

The question of a relationship between climate and cultural practices is part of the history of ideas and human civilization (Dove 2014). Theories about people's engagement with their environment are often focused on a nature versus culture dichotomy, which has been central to anthropology from its early history. Climate theory evolved from environmental determinism, arising through detailed ethnographic studies with an emic perspective, focused on material sustainability intertwined with cosmological imagination. An interest in human sustainability and the environment can be traced through anthropological history, as we see in the works of E. E. Evans-Pritchard (1940), Marcel Mauss (1950), Julian Steward (1955), and Franz Boas (1964). It is not my intention in this book to revise these classic theses. I wish only to acknowledge their significance in both a theoretical and ethnographic sense. We are witnessing new directions in climate anthropology, and new trends where climate ethnography and socioecology share an interest in an experiential understanding of people in their weather worlds (Anderson 2011; Berkes 2018; Crate 2021; Ingold 2000; Orlove et al. 2008). The New

Ecology Paradigm (see, e.g., Anderson 2011), which draws its perspective from methods assessing environmental values and beliefs, is based on a conceptual understanding of human behavior in the context of the larger environment in which all species live.

Anthropologists who are working with communities directly affected by rapid climate change have observed the problem of its denial. Some have tackled this negative mode of thought by conceptualizing it as being based on a cultural model that is evident in modes of communication. Other scholars have emphasized environmental skepticism, rejecting local-knowledge models as being applicable to climate change analyses. Yet, overall, ethnographic studies of rural communities that are most affected by climate change reveal a disconnect between people in these small communities and external authorities, with the latter claiming power over environmental scientific knowledge. This leaves us with an incomplete analysis of how these communities locally utilize their ever-evolving, place-based ecological knowledge (Marino 2015; Norgaard 2011; Paolisso et al. 2006). As Fikret Berkes (2015) notes, local rejection of climate change can be explained by looking at how local ways of knowing develop, which is through the process of accumulating and adapting knowledge, as it is embraced by the local community, as a cumulative historical experience. In contrast, many scientific observations about climate change are based on a Western positivist tradition and acceptance of institutionalized information.

An existential challenge to all humans will be more effectively managed if climate mitigation policies can be developed in response to local understandings of the social and material worlds, based on their cultural schema (Paolisso et al. 2006; Roscoe 2014:543). As a result of reviewing recent, highly influential texts on new developments in socioecology and changing climates (Crate and Nuttall 2016; Fiske 2016; Harvey and Perry 2015; Marino 2015; Paolisso et al. 2006; Petryna 2022), my writing is informed by a diverse body of literature in socioecological anthropology and phenomenology, island studies, fisheries studies, and the anthropology of place. Tracing practices and contexts involving inhabitants of particular weather worlds—specifically, those in small island communities—I have written this book in conversation with other anthropologists and scholars who bring diverse perspectives on the complexities of coastal ecosystems, also known as *blue ecosystems*, that are affected by climate change. I agree with these scholars, who emphasize that without engagement with local cultural models and local perceptions of environment and climate, local populations will reject climate policies. In my analysis, what connects these multiple modalities is the ecological phenomenological perspective I use in telling this story (Ingold 2000; Jackson and Piette 2015; Ram and Houston 2015; Van Heekeren 2012).

It is from this perspective that I argue that an analysis of the human dimensions of climate change must be reframed. In my view, anthropological and other social science studies of people affected by climate change must move beyond the position of local denial if the gap in those polarized perspectives is to be bridged. *We Live in the Water* remedies a lacuna in the literature by examining how people in a small island community internalize environmentally driven changes in both their individual and collective lives. I examine Smith Islanders' ecological knowledge, which is embodied in ordinary, everyday practices as well as in various yearly social events. This book reveals the previously undocumented manner of *being in the weather world* and introduces readers to how the islanders' ways of knowing and of *being in place* are now challenged by the politics of climate change (Ingold 2000). It examines their life narratives, which I recorded as they viewed old photographs and related stories about their work, their kin, and their life on the water, and as they tried to make sense of the rapid losses of their land and people. By documenting the consequences of their lost way of life, I conceptualize how conditions on Smith Island are shared with other global communities that are also affected by climate change.

In my discussion about people's perceptions and experiences, lived through both the material world and their cosmological imagination, I explore subjective sensory, ethical, and political dimensions. I show the human connectivity between one's sense of self and place. Examining the structures of experiences in a particular place, my work resonates with the phenomenological perspectives of Kalpana Ram and Christopher Houston (2015), Michael Lambek (1981), Deborah Van Heekeren (2012), Michael Jackson and Albert Piette (2015), and other scholars, all of whom are inspired by existential and phenomenological philosophy. I examine human consciousness, expressed in verbal and nonverbal modes of communication, in relation to a lived sense of loss, hope, and poetics on an island exposed to changing patterns.

Small island ecosystems, also known as *personal ecosystems* (Berkes 2018), are a contemporary topic of interest, as islanders often see environmental limitations more clearly than mainlanders do. In most of the recent theorizing in island studies, scholars clearly show how "islandness" has become a key concept of the Anthropocene—that is, a new epoch of humans' vulnerable relationship with nature affected by climate change (Pugh and Chandler 2021). These authors argue that working with islands in the light of changing climates has shifted island scholarship from being marginalized and romanticized geographies, instead proposing a new, alternative perspective on ways of being (an ontological question) and knowing (an epistemological perspective). I have given a lot of thought to the Smith Islanders' ways of being in place. Just as with many other small islands, Smith Island's ecology has been affected by a changing climate. Rising sea levels and seasonal storms threaten the islanders with flooding, resulting in long-term land erosion. Because traditional fishing work in this community is diminishing, many of the residents have moved to the mainland, leaving behind a mostly older population. Smith Island is a clear example of a community being transformed—indeed, being forced into extinction—although many of the islanders still hold on to the hope that their water-based lives will continue well into the future. Smith Island presents an interesting intersection of locally specific patterns of ways of being and a newly emerging, multigenerational dilemma in a small coastal community affected by socioecological changes. In this predominantly aging community, life history narratives, casual conversations, ethnographic observations, storytelling, and rituals provided me with deeper insights into the meaning of work, kinship, and community as they are integrated into the island's weather world.

Current literature addressing climate change migration shows us that many people want to continue living in an area devastated by an environmental crisis or social conflicts (Marino 2015). Thus we are left with questions about what motivates people to stay in places affected by global climate change and how they endure precarious futures. In the case of Smith Island, as the crisis after Hurricane Sandy shows, the question of relocation is an underlying threat to the community. It is important to consider that people like Smith Islanders are not part of a homogeneous community. While some leave their places of origin, others resist and are determined to endure the ecological changes. They unite their voices in a time of crisis, but on a day-to-day basis, people on the island have diverse needs, depending on their family's economic circumstance or their personal health conditions. Therefore, to address this enigmatic question, I bring to the forefront individual life histories from this community. In my narratives, I discuss the ways of being on Smith Island, or *island ways* (the islanders' term for place-specific practices). Anthropologist Michael Jackson, in a different ethnographic context, addresses such a dilemma: "That our lives confront us with inescapable quandaries and contradictions is a recurring motif in existential thought. We are born into a world we did not choose and are bound to die. Periodic crises confound our most cherished beliefs and destroy our best-laid plans. We build valuable lives, only to see our handiwork destroyed by war or natural disaster. Though our well-being depends on secure attachments, the course of our lives is punctuated by traumatic separations and losses" (Jackson and Piette 2015:156). Jackson goes on to say that if we endure, because there is always more than our individual survival at stake, our lives are always interwoven with those of significant others. The struggle for being is never only applicable to ourselves. Rather, it is "to be with others, to be there for them, to find ourselves through them" (Jackson and Piette 2015:156).

In this book, as I shed new light on how the weather-dependent Smith Islanders live with the reality of an ever-present sense of loss, I do recognize the presence of hope, as I see it, in people's engagement with their families and community—that is, in their environment. To address the human dimensions of changing ecology on this island, I discuss how, through a process of placemaking, Smith Islanders construct their social histories and reinvent their hope over generations. I regard social life on Smith Island as embodied within the larger ecological system. Most of the people on Smith Island are elderly, a common trait in many traditional communities affected by climate change. Consequently, the main threads in the fabric of my ethnographic narrative weave seamlessly together with their stories of aging. Negotiating one's older adult years depends on the fabric of cultural values, practices, and human social relations, as well as on humans' interactions with their environment.

Observing the aging experiences of islanders who, unlike their parents and grandparents, find themselves navigating this process in the absence of the next generation, while also experiencing environmental crises, requires being attuned not only to the islanders' sense of loss, but also to their skills in reinventing ways of aging while remaining in this community. I examine how people, in their later years, redefine the future hope of their community. As I tell this story, I ask the following questions. What does it mean to grow old in the Smith Island weather world? How do Smith Islanders constitute hope for their everlastingly precarious future? What are some of the conflicting tales within the overall story of this island?

To address these questions, I examine my ethnographic material from the perspective of socioecological phenomenology and the life course. I consider multiple modes of place-based knowledge, manifested in both individual and collective ways of experiencing, and I invite the reader to think in new ways about how people maintain hope through their engagement with their weather world. I take *hope* to be a mode of engagement attuned to an uncertain and precarious world. Hope, seen as an awareness of openness and promise, is relevant to a creative and critical engagement with present environmental constraints, including multiple forms of ecological loss (people, land, animals, and plants). I agree with Teresa Shewry (2015:15) that hope does emerge in contexts where environments are damaged, or where its future is denied, and argue that it is through worldwide human engagement with the environment that people draw confidence in future hope. Hope matters in communities that are affected by climate change. In the face of environmental destruction, hope is "an opening of will" (Tsing 2015:265) and involves expressions in people's thoughts, emotions, and actions. Yet, because hope is an uncertain attempt to understand experiences that are unstable, the concept of hope can't be fully settled in an absolute way, as it always coexists with loss, pain, or sorrow. Considering that environmental loss is also embodied in the future, hope is inseparable from an awareness of risk and danger as it evolves into the uncertain future.

Family Frames: The Poetics of Multisensory Ethnography

One summer night in 2015, I was sitting at the water's edge in the company of two women. While each was born on the mainland, marriage had led both to live on the island full time. As they talked about island life, one of them said to me, "It's interesting to see from this place what people take when they must evacuate the island before a hurricane. I have observed that family albums are among the many things that people often take with them." That evening's conversation on the waterfront led me toward a further interest in Smith Island family photographs and albums.

It was in Jennings and Edwina's house, in the summer of 2017, when I first encountered old Smith Island photographs. Jennings was resting in an armchair with his eyes closed when I walked into the house. He raised his head and greeted me with a smile. "It's not going any better," he said, as his smile moved to one corner of his lips and stiffened, his sadness evident in the deepening of his voice. I walked from the door to the sofa and sat down across from his armchair. Jennings, who was retired from working on the water, was a respected elder on the island and known as a storyteller. He used to be a public voice of the Smith Island community. His extensive knowledge of local history and folktales, plus his talent as a storyteller prompted Jennings to welcome reporters and researchers from the mainland. He was a founder, editor, and writer for a local newspaper, the *Smith Island Times*, where he published local news, short nonfiction stories, folktales, and poetry. He also kept an archive of newspaper clippings and family albums. Although once active, Jennings now suffered from a loss of equilibrium, which confines him to his house. His physical strength had been diminishing, yet when his wife Edwina brought out their old family albums from the garden storage house, his eyes sparkled and he jokingly said, "Oh, I don't know, I'm getting too old to say something about this."

When I pulled the albums out of the box and opened the first one, Jennings immediately focused on the old photographs. As he turned the pages, he identified all of the kids in each photo, recalling their stories. Then he stopped at one and said, with sadness in his voice, "Char and I are the last ones from our class," referring to his neighbor Charlotte. "I remember, one day Char stood up in class and said, 'I'm hungry and I'm going home,' and then she just left the class. There was nothing that the teacher could do," laughed Jennings as he finished this story. But then, as he paused, his face reflected a struggle to control the emotions this picture evoked. "Char and I, we're the last survivors from our class still inhabiting our home island," he said with tears in his eyes.

Jennings and Edwina's albums seem different from the usual family photo ones. While each photograph was systematically labeled with names and years, the overall sequences of the photographs in the albums were nonlinear. The oldest photographs, from the nineteenth century, were mixed in with photos spanning the different technical and visual styles of the twentieth century. In addition, some pages displayed old newspaper clippings discussing Smith Island, postcards sent by Smith Islanders traveling on vacations, or popular seasonal cards that included family photos.

As Jennings and Edwina shared stories from the albums' photographs—essentially *reading* their photographs—I realized that their laughter and their sadness brought joy to the process of remembering. Through their pictorial storytelling, they made old photographs, whether archived in albums, boxes, or frames, an active space, holding memories that interface with their individual and social identities. It was then, as I photographed the family albums, scanned individual photos, and recorded the narratives of other islanders, that I realized how powerful a role family photographs play in the islanders' sense of shared belonging. As Cecile, one of the island women, confirmed, "These are your memories, and if you ain't got your memories, you ain't got nothing." Later Jerry, another islander, said, "If the whole house would burn down in the fire, I would prefer these pictures over the house, because these photos are my family. I had a huge family, one time, but now they're mostly gone."

Reading photographs related to what became my ethnographic Family Frames project alongside Smith Islanders presented me with new insights into ways of being in the island's weather world. Many hidden stories resurfaced as Jennings and Edwina, as well as others, read their photographs. It was then that I began to see how collective identities overlap with individual experiences and memories, and how Smith Islanders construct their sense of self in place. As I thought about these family albums and photographs, it occurred to me that a film about the islanders' pictorial heritage, encompassing narratives about their family photos, would provide a new opportunity to recover some lost stories about people and their places. It would also give new meaning to my way of connecting people on the island with their relatives and with people on the mainland. Producing the film became my way of engaging this community with their pictorial heritage and sharing it with a larger general audience. The significance of the Family Frames project came to light during the Smith Island Art Festival, when my presentation of old photographs provoked further collective storytelling. The night before the festival, when a small group of the islanders gathered to see the material I had collected, they shared new stories. That evening, and the next day at the festival, I was surprised by an unexpected wealth of collective storytelling related to family photographs. People recognized recorded names on old photographs and linked these with faces and stories already circulating on the island and beyond. This event changed my perception of Smith Island family albums. I now understand that this pictorial archive is an available social space, one where the stories circulate in the collective consciousness of the community. When these experiences are shared, island people are left with a sense of future hope, derived from coming together and reading historical family photographs.

On a Friday evening in May 2018, the day before the Smith Island Art Festival, a small group of islanders gathered in the basement of the church in Ewell to preview my film, the outcome of a year's fieldwork, when I had collected and scanned photographs and recorded relevant narratives from twenty-six islanders. This audience responded to my film by telling other lost or forgotten stories that now resurfaced, recalled by the photographs from the past. The next day, after the premiere of *Family Frames* before a larger audience, Smith Islanders and art festival visitors engaged in another round of collective storytelling, which provoked palpable excitement among the participants. I realized that *Family Frames* not only brings to the forefront forgotten biographies, hidden in family albums, but it also invites Smith Islanders and their kin from the mainland to engage in storytelling in public. Art festival visitors witnessed how the Smith Islanders collectively entered the process of remembering, opening a new social space in which they and their kin can not only express their knowledge of a lost past, but also connect that past with their shared hopes for the island's future. This collective look at old family frames proved to be an interesting experience for all participants, showing how, through their engagement with a material object like a photograph, people create a new social space when they tell image-related stories.

Family Frames explores the materiality of the natural world and Smith Island's social histories, inscribed in the materiality of family photographs. This film, made between 2017 and 2019, is the result of a collaboration with Smith Island residents, the Maryland Public Library, and the Smith Island Cultural Center. It reflects a multisensory ethnographic participation, a practice well outlined by Sara Pink (2009). The discovery of photography in the nineteenth century was closely connected with magic. Louis Daguerre, one of photography's pioneers, was also a magician. He was searching for media that would enhance his performances on the stage of his theatre in Paris. In the process, he discovered the precursor of photography, today known as a daguerreotype. Daguerre's onstage participation in the "magical processes" (Greenwood 2020) of an evolving photographic media engaged public audiences in a collective magic. Similarly, by reading stories embedded in old family photographs, Smith Islanders, in the company of others, participate in what we may consider collective magic.

The Smith Islanders who let me see their family albums—their *family frames*—invited me into their homes and talked about their photographs. Storytelling, just like other forms of learning, plays a significant role within the Smith Island community. Following the poetics and politics in their storytelling, I have woven narratives related to the photographs from a Smith Islander's family album into each chapter in this book. As I listened to their stories and conversations related to specific images, I realized that to them, these family frames have a deeply intimate meaning, as well as social relevance. I could see the same specific image in albums, bags, or boxes from different family archives. When visiting Smith Island homes, I noticed framed family portraits displayed on the walls, on furniture, or as freestanding objects in the interiors of their houses. Framed family photographs, whether hung on the walls of various rooms throughout the home or kept in albums and boxes, seemed to clearly be more than decorative objects. They represent a family's pictorial lineage. In the absence of their children and grandchildren, many of whom now live on the mainland, the remaining, aging Smith Islanders are surrounded by photographs of their kin.

Drawing from the islanders' narratives, which they recited when reading old photographs from their family albums, I tell a pictorial story about socioecological change through their social biographies. In their intimacy and immediacy, these public and private Smith Island historical photographs encourage us to think about the relationship between appearance and meaning in the social biographies of place. The family photographs contextualized in the stories of Jennings, Edwina, and others gain substance and offer promising insights into a collective pictorial archive. I realized that Smith Islanders have developed "their ways of seeing" (Edwards 2001), looking at the world in a unique way. The photographs hold memories and provide clues to the personalities of their predecessors, while connecting these present-day individuals with their social identities.

The art of creating handmade family albums is fading. Like climate's effects on the land, photographs are being eroded by the internet. Making these historical photographs visible to a wide audience transforms a marginal archive into a dynamic pictorial social space, connecting Smith Island with the mainland. For those who moved away from the island, my film provided a new opportunity to maintain their connection with a lost home community. More stories were added on social media, and *Family Frames* is now a virtual place for kin who no longer "live in the water." In their intimacy and immediacy, old photographs, whether public or private, encourage us to reflect on the social biographies inscribed on the faces of previous generations, even as new questions about the island's future are raised. The material form of their traditional photographic heritage has changed, and Smith Islanders, like others on the mainland, now share their family frames on social media. This pictorial flow between the island and the mainland, facilitated by the internet, is a reminder that Smith Island's futures depend on an evolving mobility, what Jennings called a "change of flow." Such technologically induced change raises a new set of questions about heritage and future hopes. In this book, I argue that the Smith Islanders' traditional heritage is not a set of practices passed on in the form with which they were constituted in the past, but rather is the dynamic fluid energy of human agency, engaging in the creative process of reinventing one's current way of being in weather world, despite its diminishing vitality. From their pictorial narratives, it is evident how, through the process of placemaking, Smith Islanders construct their social history and their social identities.

The family photographs and albums I encountered during my fieldwork are an integral part of the process of "making biographies" (Edwards 2001). Visual knowledge, embodied in the sociobiographies of Smith Islanders' family albums, opens a new space for reading stories inscribed within the pictorial frames. While such albums are typically understood to be intimate pictorial spaces within the confines of kinship, in a small kin-based community like Smith Island, these private and public visual biographies overlap. In both a material and a symbolic sense, family photographs from Smith Island proved to be an interesting connector between the people who stayed on the island and those who now live on the mainland. Some families keep their albums with them on the island, while other aging Smith Islanders passed on their family albums, the frame of their family, to their mainland-dwelling children. In this book, pictorial storytelling, which evolved during my Family Frames project, is an integrated visual methodology woven into the final ethnographic narratives. It adds another, deeper perspective and shows how the Smith Islanders' senses of self and place are intimately bound together.

Ethnography is a creative process. Just like a work of art, it involves more than style, form, or playful experimentation. For me, ethnography is "the manifestation of phenomenology that rests on a grasping of what is," as Deborah Van Heekeren (2012) puts it. Elizabeth Edwards (2001:99) reminds us that in photography, as in other visual representations, the agency of the images, supported by their structure, is always informed by "a local visual knowledge" in which the viewers project meanings through their own cultural experiences. Many ethnographic methodologies—whether written, visual, or performed—are often detailed descriptions of people's experiences, knowledge, and feelings. These all contribute to human motivations and help us understand people's connections to places on deeper level. Photography has always been part of my ethnographic fieldwork. In fact, practicing photography predates my anthropological journey. Yet this is the first time that I include family photographs as an object of my inquiry, in both a material and a symbolic sense. Photography became a prime vehicle into the weather world on Smith Island and, therefore, a phenomenological mode of inquiry. I agree with Deborah Van Heekeren (2012) in believing that, as anthropologists, we can create worlds for readers, and that ethnographies are representations of mediated worlds that will extend the boundaries of comprehension.

The bulk of the ethnography in this book emerged from my engagement with Smith Island people during my fieldwork, which resulted in the film *Family Frames*. I conceptualize this pictorial archive as an active and available social space for Smith Islanders, one they share both on the island and with relatives on the mainland. The material forms in which photographs are made, presented, and viewed are not only integral to their "phenomenological engagement" and "structuring visual knowledge," but are also embodied in social modes of viewing (Desjarlais 2015; Edwards 2001; Pink 2006).

The Book Itself
Its Origins

We Live in the Water is based on my fieldwork between 2014 and 2022, when I regularly stayed on the island at various times, usually over a long weekend or for a full week. During several summers I was able to stay for eight entire weeks. This perspective is the basis for my proposal that anthropological studies of those affected by climate change should move beyond denialism to the recognition of locals as a valid and important voice in expressing the flexible nature of their shared knowledge, which we must explore more closely. Rainfall, humidity, and wind are all indicators of how much climate matters on Smith Island. Therefore, the question that comes to the forefront is how current, rapid climate change impacts these islanders.

My first fieldwork on Smith Island took place in 2014. Inspired by a brief weekend visit in 1995, I returned to Smith Island in 2013 and began my fieldwork in earnest the following year. It was on that first memorable journey to Smith Island in 1995 that I met Ken and Iris on the dock in Crisfield, when they welcomed us (my husband and me) onto their boat. Through them, in 2013 I met other people on the island. Iris was 89 and Ken was 88 that summer, and they were the oldest people on Smith Island. My conversations with Ken and Iris led to my interest in getting to know more about what growing old means on the island. I came with an open mind in my pursuit of ethnographic knowledge about the lives of people being in place. Because most people on Smith Island are in their later years, the main threads in the fabric of my ethnographic narrative weave themselves into stories about experiences in aging. Negotiating one's older adult years depends on the fabric of cultural values, practices, and human social relations, as well as on interactions with the environment. I began asking about the experiences of islanders who, unlike their parents and grandparents, find themselves navigating the aging process in the absence of the next generation. In addressing this phenomenon in the context of changing socioecology, I discuss how, through a process of placemaking, Smith Islanders construct their social histories and reinvent their traditions over generations. Thus I can offer a new perception regarding their views of ecology and their engagement with the weather world they have inherited.

My ideas in this book have emerged from numerous ongoing conversations, participation in social events, and my own sensory experiences when I stayed on Smith Island. Building on the stories told by islanders through their photographs and my participatory observations, I discuss how the water and weather define multiple aspects of daily life: prosperity, loss, happiness, joy, health, fatigue, and death. I show how the life cycle on the island, taking place within a kin-based community and surrounded by seascape, is uncertain and vulnerable. But, at the same time, it offers a deeply satisfying habitat for its populace. Many of the ideas I present were sparked by conversations with the residents during days spent on the island, followed by speaking to my students about the anthropology of climate change.

The ethnography of placemaking presents an opportunity to address a special conception of the history of Smith Island. It shows us how, even when stripped of its future, a place can give people a new sense of the past and of themselves as it intertwines with their sense of place. While Smith Islanders notice every detail of their changing landscape, when they are telling stories and reading old photographs, their relationship with the land also keeps them connected to their ancestors through the process of reinventing their cultural traditions. Between 2017 and 2019, in addition to exploring pictorial histories, recording structured interviews, and engaging in casual conversations with individual islanders, I participated in social events, such as church meetings and traditional annual celebrations on the island. Smith Island's oral histories, family photographs, and social events provide not only the sociocultural context for this book, but underscore how life on the island is uncertain and vulnerable, yet where they live is also a deeply satisfying habitat to many.

In the chapters of this book, I have organized the ethnographic material to show how the processes of being in place and placemaking are intertwined. As I discuss the islanders' knowledge of the material world, as well as their abstract ideas and moral values, I emphasize how their experiences are embodied in their socioecology. I focus on a small community with a specific island socioecology and tell their story about future hope through ethnographic narratives, as well as through less traditional methods of pictorial biographies, told in photographs and albums, as read by Smith Island people.

Its General Outline

In chapter 1, "Weather Is Everything," I establish the fundamental concepts for my analysis of Smith Island's socioecology in an environmental context, weaving together weather, water, place, and community. I discuss how these, plus the Smith Islanders' other modes of knowledge and their belief system, are connected to the process of placemaking in their inherited weather world. I emphasize how these islanders' cumulative body of knowledge is shaped by their collective experiences and observations, which are brought together, interpreted, and shared through storytelling. While exploring how people mingle in this kin-based community, I show how imaginative systems, such as beliefs and dreams, and ecological systems overlap when framing people's views of climate. Exploring these themes, I take the reader into the home of one of the oldest Smith Island couples, Jennings and Edwina, as they share stories arising from handwritten notes on the backs of old family photographs and recall memories from their decades of living in the water. I illustrate how many of their stories include elements of local ecology and weather, such as how, at sea level, even the smallest amount of rain can cause dramatic flooding in certain low-lying areas, slowly reducing the landmass over time and significantly reshaping daily life. As water is constantly in motion with the rhythm of the ocean, the pull of the moon, and the push of the winds, Smith Islanders have learned to mirror these forces in their own movements, having discovered that living in harmony with nature is key to their survival.

This intimate conversation serves as an outline for my work, as it underscores the islanders' own perceptions of living in the water, firmly rooted in their complex knowledge of Smith Island's dynamic ecology. I make the point that an island-based study enables us to think differently about human knowledge and climatic futures. The motions of wind and water evoke powerful emotions, shape the way islanders view their role in the cosmos, and are evident in the very language they use in forecasting the weather. Sensory knowledge of the wind and their traditional way of reading the sky show the islanders' agency, expressed in their response to post-hurricane recovery. Recognizing the ever-increasing, precarious nature of living in the water, Smith Islanders have begun to develop new economic strategies and traditions projected to sustain their way of life.

In chapter 2, "Ways of Knowing," I explore the process of becoming a waterman, a way of life passed from one generation to the next, by focusing on the life narratives of a Smith Island waterman, Edward, known as "Eddie Boy." By examining Eddie Boy's stories across his lifespan, I reveal the power of kinship and experiential learning. I focus on the dilemma of aging in a changing environment, one that Eddie Boy shares not only with other Smith Islanders, but also with people in small rural communities worldwide. As a community ages, stories about people and places often become lost as future generations dwindle in number, often moving away.

I explore this phenomenon, this "lost language of the watermen," and what it symbolizes to Eddie Boy, relating it to how working within the environment defines one's sense of self in a small kinship-based community. Lastly, I examine how connections between the environment, work, and one's identity are integrated into socioecological systems.

In chapter 3, "Land and Water," I explore Smith Islanders' engagement with their land. Historically, this land has been managed by families caring for the properties surrounding their homes. Despite community-built channels easing the connections between settlements, the growing problem of the land's rapid erosion has forced islanders to adopt new approaches to its management. Because these people have historically been able to manage their own land, they feel as if the current problem will be solved in the same way as in the past, although the recent rate of erosion is causing an existential threat to the very island itself. Reflecting on my conversation with Bobby and other residents, in this chapter I argue that their sense of self is embodied in the multitude of their engagements with the island's grounds. Bobby is a family man in his forties who, like many islanders, notices the slightest changes in the land and climate and always processes his observations into how best to engage with the environment. I then turn to the subjects of self and place as I relate the sensory experience of exploring the landscape, and I discuss the integration of ecological knowledge into land management practices.

Chapter 4, "Shifting Grounds," examines changing socioecology through Smith Islanders' relationship with death. One challenge they face comes from increased longevity, as more people tend to live into old age. I examine this topic through the lens of Ken and Iris's experiences. They are a couple in their nineties who have weathered their share of illness and death. Focusing first on their aging trajectories, I offer insight into a slow process of losing the patterns of traditional island ways late in one's life and bring attention to new questions about the parallels between social and environmental death. The second part of this chapter examines how a collective grief over the social losses islanders must endure is expressed through the ritual structure and poetics of storytelling, joking, music, and poetry. I focus on the social aspects of the rituals surrounding a death, the common expressions of grief, and the unique way Smith Islanders view the passing of loved ones. For these islanders, the sense of loss associated with death is embodied in their language, land, and way of life. Focusing on the process of collective grieving, I describe a Smith Island funeral and examine how collective grief is shared through the structure of ritualized collective gift giving.

Chapter 5, "Broken Bodies," discusses how the environment affects human health. Most people on Smith Island are in their later years. Therefore, health and illness are a focal point of their everyday experiences. Increased longevity presents new health challenges for residents who work on the water all year. The seascape environment is hard on the human body, and many suffer chronic pain in silence, but when an islander has a terminal illness, others in this community show compassion for their neighbors. By presenting Mary Ruth's narratives of a brief history of health care on the island, I show how healing and medical care are found not only in modern and traditional treatments but also in the spiritual healing practices of the church. In the second part of this chapter, I turn to Kathy, a newcomer who suffers from post-traumatic stress disorder (PTSD) and moved to the island to find a peaceful place for her mental well-being. Kathy speaks about her belief in the curative agency of Smith Island's natural environment and emphasizes the healing power of the collectively expressed compassion among people living there. Peggy, who lives alone in her later years, cares for others in pain, even when facing her own illnesses, and shows how the subjective and social dimensions of health are intertwined. I show the power of collective bonding as I discuss the social dimensions of health and illness on two levels: as a constant source of uncertainty, and as a socially shared intimacy. The uncertain nature of the weather and the sea mimics the same uncertainty that islanders feel toward death. Stories of the dead are conveyed through the filters of place and weather, as well as through the lens of sudden, tragic illness. Smith Island, with a society composed mostly of older adults, faces the major health challenges of a population in decline. The pain and hardships caused by illness are not private matters to these islanders, but instead are a source of collective community concern. News of illness travels quickly through the local grapevine, relayed through conversation, social media, and casual exchanges between islanders as they greet one another in passing. Personal health issues are even made public during communal prayer services at church and other social gatherings. This lack of privacy, as some would characterize it, takes second place to the extreme hardship brought on by illness. It serves to bind members of the community more closely, as living in the water presents its own unique set of challenges: long boat rides, travel on the mainland, overnight lodging, and other unexpected costs of procuring medical help.

In chapter 6, "The Taste of Things and Comic Relief," I show how one of the most distinctive features of Smith Island society is the gender specificity that the weather world imposes on its socioeconomic structure. Men are defined locally by the traditional masculine work they do on the water, while women are multitaskers, constantly reinventing a wide spectrum of skills that are vital both to their families and to the community at large. Watermen develop multiple skills to sustain their crabbing and oystering, yet women's work on the island has been historically even more diverse than men's. In this chapter I discuss this division of labor by gender as it relates to life in the water and to how it lends itself to a certain social cohesion between members of the same sex. I focus on this gender cohesion as I explore the rituals of Smith Island women, framing my discussion around the annual event known as the Ladies Dinner.

Despite social hardships and environmental difficulties, this sense of cohesion within genders also extends to the Smith Island community as a whole and is reflected in the major traditional events shared by all islanders. Every year the women engage in organizing these celebrations, which reflect the collective bonding among neighbors as they gather to share food, song, prayer, and humor and unveil their common love of music and playful storytelling. This chapter shows how deeply islanders embrace this collective life and reveals their strategies for reinventing the rituals and social bonds from their heritage in the face of changing socioecology.

Chapter 7, "The Art of Creating Futures," discusses some of the trajectories of newcomers and their participation in constructing Smith Island's future. Despite all their losses, the remaining long-term Smith Islanders are moving forward with their daily tasks to secure their survival and, together with the newcomers, build resilience by reinventing the island's heritage. In this vein, I turn to the life stories of the newcomers, who have distinctly different backgrounds and motivations for moving to this island. In discussing how Aram, Steve, Shawn, Pamela, and Kathy became part of the Smith Island community. I show how the process of their integration changed the focus of their lives, as well as how their transformations are embodied in the imaginative process of shaping Smith Island's future. I describe how they became part of the community in their own unique ways. They alternate between feelings of belonging and alienation, yet they also experience a sense of the magic they find on this island. Examining the complicated process of reinventing the island's heritage, particularly from the point of view of a newcomer, provides additional new thoughts about the implications of how climate impacts humans' conditions, heritage, and creativity in response to socioecological change.

Its General Purpose

We Live in the Water offers a new anthropological look at the human dimensions of climate and weather from a multigenerational approach, addressing both personal and communal future hopes. I argue that a life-cycle perspective reveals a need for scholarly attention to the interrelationships between changing socioecology, collective rituals, and the environment. I examine Smith Islanders' agency as they reinvent their individual and collective ways of being in the seascape's weather world. The purpose of this book is to contribute to a better understanding of growing old in a small island community that is affected by a changing ecology. In it, I examine a process of placemaking, looking at how Smith Islanders construct their social histories, as well as reinvent their traditions. I discuss how aging presents its residents with new dilemmas and demands. Linking the concept of people's traditional ecological knowledge to the shifting demands of their surroundings provided me with tools to illustrate how the processes of growing old alongside environmentally driven changes affects a small island community.

Exploring how Smith Islanders developed confidence through their understanding of uncertain weather, and how they have mastered and enjoy their present way of life, regardless of its precarious nature, I argue that their art of being in the weather world is a dynamic and creative collective process to achieve liminal sustainability in what can be seen as the light or the shadow of the precarious. After being with Smith Islanders for over a decade, I view their subjective experiences as being inseparable from their collective life. Using these subjective experiences, I show how the islanders' sense of self is connected to place. Considering various forms of social life on Smith Island sheds light on the islander's efforts to reinvent their heritage and reveals mutations in many of their traditional practices. During their *socials*—a term the residents use to refer to social events held on the island—Smith Islanders renew their collective bonding and refine the shared voice that will represent their way of being to the outside world.

1

Weather Is Everything

JENNINGS EVANS

In the Water

"We are different here," Jennings said while reading the island life inscribed in old photographs. Then he paused, and the twinkle in his eyes quickly dissolved as he focused on his next thought.

Water makes a difference. We are not on the water, but we are in the water. That is a big difference. Weather is everything here. We depend on weather. Weather is driving our lives. You got to respect weather. We live in the Bay, not on the Bay. We have been here hundreds of years and survived hurricanes and we are still here. But we don't know what will come with the future. We have to move with a different pace. You go out and you don't know if you will come back. It is too unpredictable. Water is a blessing and a problem at the same time.

Defining ways in which Smith Islanders tackle the ever-changing and uncertain weather world they inherited from their ancestors, Jennings's words offer an insight into the Smith Islanders' underlying perceptions of their "being in the weather world" (Ingold 2011:73). In this chapter, I follow Jennings's narratives about Smith Islanders' modes of knowledge and belief systems as they connect with the process of placemaking in their inherited weather world. I emphasize how their cumulative body of knowledge is shaped by their collective experiences and observations, which are brought together, interpreted, and shared in storytelling. While exploring how people mingle, both in this kin-based community and in their environment, I show how imaginative systems, such as beliefs and dreams, and ecological systems overlap when framing people's views of ecology.

The water moves constantly with the rhythm of the ocean, governed by gravity as well as by changing winds, and Smith Islanders have learned to mirror that movement in their own actions. The tides—determined by the cycle of the moon and the rotation of the earth, and influenced by wind direction—have a significant impact on the work and life of those who make their homes near the shoreline. Because Smith Island lies at sea level, even the smallest amount of rain contributes to higher water accumulation in some parts of the island and thus reduces its landmass. Water is a blessing, because Smith Islanders utilize it to sustain their livelihoods, but it also is a challenge, because the constant movement of the water continuously alters the ground and significantly erodes the land. The art of living on Smith Island, "in the water," derives from the islanders' ways of balancing this aquatic *double bind*, a concept Gregory Bateson (1972:125) discusses in the context of an ecological paradigm. To be in their weather world, as Jennings has reminded us, "we must go with the flow." In other words, we must synchronize human life and the landscape/seascape rhythm, as best we can, in flexible and fluid ways. From Jennings's narratives, we can see how dynamic movement is one of the habitual underlying conditions Smith Islanders endure, a part of their place-based knowledge. Therefore, I posit that their traditional ecological knowledge is a guiding principle of their flexible engagement with the environment.

In Jennings's narratives, we are faced with an interesting example of how people construct their senses of self in the *blue spaces*, a term used in aquatic scholarship (Foley et al. 2019; Hastrup and Hastrup, 2016; Strang, 2004, 2015). Jennings's expressions—"*weather* is everything here" and "*water* drives our lives"—define the habitat within which Smith Islanders construct their lives. It is closely related to the fundamental question about a sense of self and place that I address in this book. As rising ocean waters prolong the flooding seasons, the islanders' daily lives are disrupted more often. Their life flow is subject to change, subordinated to the high tides pushed into the island by the wind. Thus Jennings's statement speaks to scholarly concerns about the human dimensions of ecological change.

Water represents a fertile force, yet a blue space often remains uncertain. It is mysterious in its depth, and its infinite horizons cannot ever be fully controlled by people. Veronica Strang (2004) explores water, in multiple forms, as a part of all living organisms when she shows how the properties of water reveal obvious links between

ALAN MARSH

macro and micro systems, and how bodies of water are transformed through a cultural lens into a range of cosmologically imagined flows of water managed by supernatural powers. While water shapes diverse sensory perceptions and meanings, it also has universal, consistent qualities that are evident in its continual change and movement: "People imagine ideas and emotions flowing between themselves and the world, and in this system, as in any other, there is a 'proper order'—boundaries and limits, a correct balance of flow" (Strang 2004:68). In cases of either extreme flooding or drought, water causes either a transformation of or an end to ecological systems (Crate 2021; Crate and Nuttall 2016; Marino 2015). Water flow affecting Smith Island is always determined by weather, as weather is by the wind.

The Wind Is the Weather

It was at the end of a hot summer evening, just after sunset. The sky was turning dark blue, and strips of orange-yellow light outlined the horizon west of Smith Island. I was walking home from the Smith Island Methodist Church's camp meeting with a woman named "Big Doll." "She is old and needs a new seat," remarked Big Doll about her bicycle, explaining why she was riding with one foot on the ground. Big Doll always travels around the island by bicycle and often carries many things on her handlebars. The evening air was hot and heavy as we started walking away from the campgrounds. Suddenly a gentle breeze began to blow from the North End. "Doll, do you feel the wind?" I asked. "Yeh," she smiled, and with joy in her voice, called out loudly, "Heaven on earth." It reminded me of Jennings's words, and I said, "The weather is everything, here on the island." Doll, still smiling, replied, "And the wind is the weather." That moment was a "rich point" (Agar 1994) in an ethnographic sense, as it offered me an insight into how weather can be emotionally experienced in a place where life itself is heavily dependent on it. Days later, on another hot evening, when walking outside and photographing, I recalled that moment when the orange light of evening illuminated the shanties and boats in Ewell's fishing marina and cast deep shadows across the landscape. I wandered by David's shanty. It was open, and his wife Cecile was *fishing up*, or tending to the floats containing soft shell crabs. I could see that Cecile was tired from the heat and a long workday. As we chatted, an unexpected breeze came from the northwestern side of the island. In the middle of her work, Cecile lifted her arms and gaily called out, "Heavenly wind." These two encounters with Doll and Cecile reminded me of how just a small gust of wind can transform summertime on the island, with humid heat trapped in the grasslands, into a joyful sensory moment. Big Doll's and Cecile's expressions of joy from the touch of the wind allow us a glimpse of how the sensory experience of the wind's flow registered in their imagination.

Historically, wind energy was vital to fishing voyages when the skipjacks were under sail, but it can also bring loss—and sometimes death. To the islanders, a cooling wind brings release, manifested in the sweet smells and familiar sounds from the island's marshlands. But in the spring and fall, wind energy is a major factor in flooding and subsequent land erosion on the island. During big storms, powerful winds may destroy houses and shanties; islanders would say, "It blows a gale." "Pray for the rain," wrote one Smith Islander on Facebook Messenger when a house on the island caught fire and wind began to spread the flames. For the islanders, wind, like the water, is both a blessing and curse. Wind, as Tim Ingold (2011, 2013) reminds us, is material energy in motion, movement between earth and sky, air currents in an inhabited weather world. In my view, lived experiences in the realities of the weather world are manifested in multiple modalities. Islanders respond to wind through their sensory experiences, which spark people's emotions: either joy evoked by a touch of wind, or fear from the wind pushing against one's body in a storm on the water. They also respond to it through their imagination, informed by their Christian belief system and their ecological knowledge. The latter manifests itself in practices like reading weather, which is an essential skill in navigating the seascape.

"I can look at the sky in the evening around sunset and tell you what kind of day it will be the next day," said Junior, one of the older watermen, when I asked him about Smith Islanders' observations of changing weather patterns. Part of reading the weather is observing the general aspect of the sky and nature. Jerry explained why observing weather in the seascape, on the water, and on the land is part of their everyday life: "Watermen are always watching both, the weather on the water and the land. When we are home, a change in strong wind will require us to move our boat. For example, if the back of the boat, the stern, would be facing south and strong wind was to hit, seas could wash over the back and cause the boat to sink. So, watermen would need to turn the boat around and have the bow of the boat facing south." Then he explained some signs related to weather and the wind: "We can read signs and know somewhat about what the weather will do. When I see sundogs around the sun it will be bad weather in two to three days. The sundog is a small length of rainbow on the left and right side of the sun. Also, when clouds are low and begin to break up into small clouds, we know a strong wind is coming. This usually happens ahead of thunderstorms."

Smith Islanders use specific words when they define a type of wind. "It is blowing a gale" is shorthand for winds that are blowing extremely hard. When there is no wind at all, it is "slick-calm," meaning that the water looks slick, not disturbed, Mark explained. Watermen call the clouds along the horizon in the direction of sunrise, with clear skies above, a *lee set*. When seen early in the morning, it is a sign of forthcoming wind. One of the watermen added, "You can be sure you are in for a blow; the wind will come that day, not next week." Mark followed up our conversation and sent me a picture of the sky from his phone, with the following note: "This is called 'sundog.' It could be on either side of the sun. If weathermen see this in the morning, they say that it is a sign of bad weather later in the day. If seen in the afternoon, it is a sign of good or clearing weather."

In addition to reading the sky, Smith Islanders observe animal behaviors in their environment that signify a change in the weather. Eddie Boy pointed out to me how he and others left a herd of goats on the neighboring island northeast of Ewell and across from his shanty. He can now make a weather forecast from the goats' movements: "I watch them goats, how they move from one end to the other and can tell what is coming in terms of weather." When people spot dolphins on the surface of the sea, they know that a storm called a *nor'easter* is on the way. This is a strong front of wind and rain that moves up the East Coast, similar to a mini–tropical storm, and may last for several days. In addition to reading weather signs, islanders

"SLICK CALM"

also follow the movements of weather patterns. For example, as Eddie Boy noted:

> The island is saved from a lot of thunderstorms. Thunderstorms travel with the tide. If the tide is falling, a thunderstorm will travel south with the tide. Same for a rising tide; it will travel north. Smith Island sits in the middle of a triangle of channels. The ship channel to the west, Kedge Strait . . . to the north, and Tangier Sound to the east. The point of this triangle is below Tangier [Island], where the Sound connects with the Bay. These channels will grab a thunderstorm and move it away from the island. This is not always the case, but it happens a lot.

Knowledge about the weather is often shared, expressed through "weather talks" (Paolisso 2003) during watermen's morning coffee socials at the store, on the waterfront in their shanties, and over marine radio when on the water. "There is always a conversation about the current or future weather conditions. Weather talk is usually part of initial greeting, and rekindled when someone else comes in," confirmed Mark. He further explained, "When it looks like a rainy day or rain is coming soon, we say, 'it looks like rain,' or 'it's breezing up,' or when wind is decreasing, we would say 'it's moderating,' and when wind is changing directions, we would say 'the wind is hauling around.'" Mark then added, "Most watermen call a barometer a 'glass,' short for weather glass, and say 'it is down,' which means that the barometric pressure is low, and therefore would mean that a change in weather usually brings wind when it starts back upward." Examining Smith Island weather talk presents an interesting opportunity to see how knowledge about short-term weather events and climate patterns articulate expressions of ever-evolving language heritage on an island, according to linguist Natalie Schilling (2017).

Junior spoke about a dark blue line and other visual signs that, for him, are indicators of the next day's weather. When I asked him to talk more about what weather and wind mean to watermen, he replied, "We are born with instincts and as we grow our parents teach us about the weather. Then we grow older, and we have an instinct to tell us about the weather. The instincts you have obtained can make you a good waterman or a failure." When I asked Junior if what he meant by "instinct" is knowledge, he responded, "Yes, it is the same, you learn by experience, you remember the storm on the water or other situations like it, and never make the same mistake, and as you are growing up you collect these experiences." He then went on to describe one of those experiences: "It was one time on my father's sailing boat, the skipjack we call it, that mast broke in the storm, and we made it across the Bay, but I will never forget."

For Junior and others of his generation, experiential learning when coming of age is clearly an important process of knowledge-building in place. Junior talked about one such instance: "When I turned sixteen, my dad sent me out on the water alone, he trusted me. It was a long difficult day, and I missed my birthday celebration, one you get on the island when turning sixteen, but I didn't mind." Junior's description about being out in the open, "on the water" in difficult weather, brings to our attention the topic of death in light of one's life course. The sense of "nearness to death" (Turner 1969; van Gennep 1961) is a part of maturing and learning how to be a successful waterman, all of which takes place on the water. Junior told me that a "successful waterman knows when to not go out on the water," indicating clearly that risking one's life on the water not only is personal but also endangers one's family, who depend on a man's work, and community members, who may have to risk their lives when called upon for help.

Weather is inseparable from Smith Islanders' experiences of being in the world. The weather world is deeply connected to their sense of self. Weather is interwoven into the patterns of their social life. Weather is expressed through the poetics of emotions and beliefs during greetings and conversations, also known as weather talk. Chris told me that the weather is always "first in the minds of Smith Islanders." It is experienced, observed, and predicted, as well as remembered and storied by them. When Jennings and his neighbors returned to the topic of being in the weather world during our conversations, they emphasized how, for Smith Islanders, weather defines their prosperity, their losses, and their feelings of happiness, fatigue, joy, suffering, and death. In this book, by drawing on my fieldwork interviews and conversations, I will further explore the process of becoming a Smith Islander. I will look not only at the experiences of growing up and growing old on the island, but also at the processes of fitting in, as well as discuss the lived experiences of newcomers who have chosen to move to the island.

God's Gift: Dirt and Crabs

Smith Island's future is precarious, like that of other islands worldwide. It is through conversations about the island and rising water, like those about the weather and wind, that the subject of religion enters discussions about climate change and sinking land: "If God wants us here, He will provide, and when it's time, He will show us what to do." Smith Islanders see themselves as survivors on the land that is given to them by God. "God give me a piece of dirt to live on, crabs in the water to catch, that is all I need to survive," go the lines from a Smith Island song. The islanders define their life's purpose through their belonging to a place and enduring the conditions they inherited. In their view, it is God's will that protects them and determines their future. God, therefore, presents an obvious focus for a time when they are facing uncertainty. In the extreme, the unpredictable conditions that are ever present in islanders' lives are embodied in a sudden death on the water or in their island home, far from the mainland's medical help. "It is like being on a boat and bad weather or an accident happens. You must hope for the best. Do what you can do with your limited resources and allow things to play out, hoping and praying that in the end all is well," explained Mark, one of the mid-generation watermen. When I asked Junior about his faith, he emphasized, "I talk to Him every day, juts like I am talking to you. It is part of my life. It is for our benefit, because I tell you one thing, God makes no mistake, there is nothing new we can make. I am very thankful for an every day I have."

For generations, survival on Smith Island has required knowledge of the Bay's ecology. Water and weather, controlled by the unpredictability of wind, are humbling forces to them. Mark's reflection brings us back to concepts of ecological knowledge that, when combined with a flexible way of being and with faith, are critically important to Smith Islanders. The limitations imposed by their integration within larger, complex ecological systems are what connect them, in their perception of a self rooted in time and place, with their ancestors and with the agency of God. Smith Islanders express their faith in God in the company of others during weekly church meetings, bible-study groups, and other social events on the island. The Methodist church is a center for communal faith-based life. Organized gatherings consist of readings from and analyses of the Bible, collective singing, and praying. Their prayers reflect gratitude for the life they have, as well as anxieties resulting from the uncertain conditions and challenges related to the environment, illness, or death, each of which poses threats to their overall prosperity and health. Prayers are informal, and, as the islanders express their emotions openly in the company of others, they share their vulnerability.

A belief in God is more than an impulse to passively persevere in being: "It consists in the search for adequate ideas that enable us to actively sustain our sense of presence and purpose" (Jackson and Piette 2015:12). I see Smith Islanders' acknowledgment of and faith in God as being fully integrated into their ecological knowledge and outlook in their weather world. Unpredictable weather, knowledge of the Bay's ecosystem, and humans' humility (manifested in a reliance on God) are interlocked and register in Jennings's and others' voices. This closely corresponds with Michael Paolisso's (2003) argument about ecology and spiritualty on Deal Island, not far from Smith Island. He shows how Chesapeake Bay watermen's ecological knowledge, a science encompassing the Bay's climate and blue crabs, is closely connected with their spiritual belief system (Paolisso 2003:71). A range of authors—including Eugene Anderson (1996), Fikret Berkes (2003), Tim Ingold (1986), and Anita Maurstad (2004)—have proffered ideas relating to the ecology/faith paradigm Gregory Bateson (1972) discusses in his publications. "They give the sea an agency; they talk of the sea as they talk of the Lord; it gives, and it takes," writes Anita Maurstad (2004:280), discussing the bond between ecology and cosmology when she describes how fishermen handle challenges presented by the North Sea. In her discussion about Christianity and magic in Melanesia, Deborah Van Heekeren (2012) shows how phenomenological questions address the art of everyday ethnography and its "grasping of what is." Tim Ingold (1987) connects mind, nature, the material world, the economic needs of humans, and one's life, all of which are imaginatively constructed in mythological stories and religion. Eugene Anderson (1996), Fikret Berkes (2018), Stephen Lansing (2006), and others argue that through their belief systems, non-Western societies manage a sustainable balance in their engagement with the environment. They also posit that, in a small-scale society, it is a profound respect for the environment and how it is treated, governed by the rules imposed on people through religion and cosmology, that sustains this balance. These authors follow in the tradition of Gregory Bateson (2005)—who argues that both the sacred and the scientific are integrated into the sum of ecological knowledge—and continue to be relevant for anthropologists concerned with people's engagements with the totality of ecological systems. Bateson (2005:88), in writing about the unity of mind and nature within complex ecological systems, states, "Since it is not an individual or even a species that is the unit of evolution and survival, but rather the encompassing system, the faith of the individual and its environment are intertwined. As a result, it is impossible to compete with and dominate, or win against one's own environment." Bateson's (1972) ecological approach to the analysis of religion and cosmological belief system has influenced not only environmental and ecological anthropologists Roy Rappaport (1979) and Leslie E. Sponsel (2012), but also others addressing religion and cosmological systems. To demonstrate such unity, in the next section I will examine how ecological knowledge is shared through stories in the Smith Island community and discuss how a story about a dream exhibits ecological knowledge, which plays an important role in finding a lost dead body, swept out to sea by a storm.

The places where people live and work, and their histories and values, have been critical in understanding how they perceive not only the weather, but also long-term climate change. When I ask about the impact of climate on Smith Island during weather conversations, some individuals speak of disappearing land, due to erosion, and some avoid climate change conversations altogether. Others, like Eddie S., talk about the influence of changing wind patterns: "Not all the watermen I know noticed sea level rise. Some made marks that indicated the volume of water was the same. What I had noticed, and Eddie Boy confirmed, that our wind patterns have changed. We used to have mostly westerly winds and now we have many more easterly winds, which has always caused us to have high tides." When I visited Smith Island between 2021 and 2022, the water level was noticeably higher, and flooding had become much more serious. During high tide, water deters people from walking across the island and, in some instances, prevents them from leaving their houses. The community works together to manage water levels by maintaining channels and road elevations.

Storied Wisdom in Dreaming and Places

"Any crisis that comes up has been dealt with years back, and that's how stories came into being," said Jennings during one of our conversations. These stories, based either on actual events or dreams and visions, represent older wisdom and are available for Smith Islanders across generations. Ethnographies of humans' engagement with their habitat tell us that people historically have projected meanings onto places (Basso 1996). Place-based stories and dreams, closely related to knowledge of the seascape, often speak about Smith Islander's relationships with the dead. "Death of any community member brings people closer together here; we are all interconnected," said Jennings. "A good crabbing season is taken day by day. It's only at the end, when the storms have missed you and the catch has been good that you relax," added Chris, a former waterman, when reflecting on anxieties about uncertain weather conditions experienced by people, like Smith Islanders, working in the seascape. Chris commented on Jennings's words: "Smith Islanders spend as much time thinking about the weather as farmers, even more so. The weather can bless you or kill you. We don't dwell on the science of weather, or climate change, but we do depend on certain signs to guide us. Those signs may have a scientific basis, but it is the implications of those signs that interest us. The signs of climate change are more subtle, so there is natural skepticism. While we are aware of our history, we think in terms of the season in which we are working."

Witnessing the death of a kinsman, at sea or on the island, is always traumatic. Yet, as Jennings pointed out, a nearness to death is always part of day-to-day life: "You go out and you don't know if you will come back." Stories about the dead are structured according to knowledge of a place, a body, or self and its relation to faith, and they often include hints of a secret. Some stories and dreams overlap, just as they do in the tale I encountered while having a conversation at the library with Anita, who spoke about her uncle's death: "He was going out in a small boat, just to check on something. Weather changed fast and the wind was so strong that his little boat was swept up. His body was lost in the sea, and they couldn't find him for days. Then Eddie Boy, one of the watermen, had a dream about the place where he was. They went there and found him." When I asked Eddie Boy about this story, he confirmed it.

> It was Sunday afternoon, after church, when I came home and, like my usual routine is, I eat and lay down on the sofa. That

day I couldn't fall asleep. I was slowly drowsing away when in my mind I keep returning to one place on the water in a grassy area. I got up and told my wife June that I must go there. I was thinking for a moment that I may go later, but then I went, and the body was there. I am glad I did, because if I waited for later, we would have never found the body.

When Eddie Boy finished this retelling, a chilling silence hung over the living room. This dream, born out of Eddie Boy's fear concerning the body of a fellow islander, detailed a vision of the seascape in the wind. In it, he reconstructed the dead man's being out in the weather world at a particular time and place, which prodded him to further action. As Eddie Boy told the story, he was now sure about the connection between waking-time events and events in his dreams. I take Eddie Boy's dream, during the light sleep of a Sunday afternoon, to be a flow of thoughts emerging from ecological knowledge and sliding across the limits of actuality into dreamtime. Jeannette Mageo (2021:3) argues that dreams are extensions of cultural realties, and the visual metaphors in dreams represent cultural schema and models, which not only determine daily concerns and their intensities, but also provide special access to lived subjectivity in a particular historical moment: "Dreams are descents into unguarded experiences—into preoccupations, questions, fears, wishes and mental wanderings."

Eddie Boy's tale, like other Smith Island stories, is retold in new contexts and circulates throughout the community. The dream story of his passage through the landscape has become part of the wisdom associated with place, "the wisdom that sits in places" as Keith Basso (1996) puts it in his ethnography about land, values, and storytelling. The practice of storytelling has been deeply established for generations in the life course of Smith Islanders. While all age groups on the island participate in storytelling, elders are the keepers of old stories, reflecting the values and knowledge of their community. "At Tangier they call us 'yarnies,' meaning we could tell long yarns [stories]," said Jennings with laughter. Storytelling relates to Smith Island life in the broadest sense. These tales are concerned with serious tasks, such as weather, work, social life, faith, or death, but they are also used to entertain. Some stories, deeply rooted in the island's oral traditions, are retold often and could be called old yarns, but each generation has its new stories to relate. Shared knowledge of living in the water is cognitive as well as sensory. It is connected with Chesapeake Bay's climate, water, crabs, fish, oysters, birds, turtles, eels, and many other animals, all of which co-inhabit the ecosystem with the people living on the Bay. On Smith Island, as stories and dreams overlap, the act of *storying dreams* is one of the modalities for learning through traditional ecological knowledge.

As watermen on Smith Island grow old, participation in storytelling becomes a creative way of connecting people and places across successive generations. Storytelling, just like experiential learning, has a significant place in the life course of people on the island. Smith Island stories are defined by membership in a closely knit, kin-based island community. This storytelling practice has served for a long time as an effective way to pass on knowledge and provide a space for elders to actively participate in the social life of the community. Stories are told during social gatherings and community events, and they are often part of casual talks.

Storytelling *is* a form of experiential learning for the Smith Island community. "Elders provide corporate memory for the group, the wisdom to interpret uncommon and unusual events and they help enforce the rules and ethical norms of the community," writes Fikret Berkes (2018:132) about storytelling later in life. "It is in [the] power of storytelling that human knowledge resides," writes Tim Ingold (2000:164) in his work about humans' ability to weave stories from the past into "the texture of present lives." I agree with Ingold's argument that knowledge is the evolving, active process of remembering, and that stories, from a knowledge point of view, do not always represent the world, although they do "trace a path through it that others can follow" (Ingold 2002:161). Making their world through stories and connecting the notion of one's soul with one's world is deeply embedded in Smith Islanders' socioecological heritage.

Barbara Myerhoff (2007:89) compares storytelling to soul-making: "Growing souls through stories is the making of the world." Because the texture of life on Smith Island is now deeply affected by rapid socioecological change, population fluctuations, and land loss, the traditional intergenerational modes of communication are challenged. In their late adulthood, Smith Islanders are holding on to their knowledge through their stories. Yet, in the absence of a next generation, they face a painful dilemma, one expressed by Dwight, one of the watermen: "There is hardly anybody here now to tell the stories to." Following the ensuing silence, he rested his voice and let his eyes wonder across an album with family photographs that he inherited from his mother.

In the past, Smith Islanders' traditional knowledge was always integrated into intergenerational relations over the life course. The process of becoming part of the island's ecology in the company of their kin has been essential for every Smith Islander's sense of self and community.

Traditional Ecological Knowledge

"We call it crabbing" Dwight told me, "because we go out, just like hunting, you look for what you can get, you must be observant of what you know and watch the patterns of crabs' movement, that is the art of hunting. But being out there and free, being your own boss and going where you please, is as important as the money you get for what you find." Smith Island watermen are known for their studious

DWIGHT MARSHALL

approaches to crabbing, and for thinking their way into the crab pot (see, e.g., Warner 1976). "If you want to catch the crab, you have to think like the crab," I heard Eddie Boy say.

All their knowledge is recorded in watermen's journals, called logbooks, and passed on by them to the kin they mentor. A Smith Islander's definition of traditional life is closely related to the concept of traditional ecological knowledge (TEK), defined as "a cumulative body of knowledge, practice, and belief by adaptive process and handed down through generations by cultural transmission about the relationship of living beings (including humans) with one another and with their environment" (Berkes 2015:7). TEK has also been well researched by other socioecologists (e.g., Berkes et al. 2003; Ingold 2000; Paolisso et al. 2006; Ponkrat and Stocker 2011; and Roscoe 2014). Trajectories of people's engagements with their weather worlds that have been documented by anthropologists show evidence of how those in weather-dependent occupations, like farmers, fishermen, and pastoralists in both Western and non-Western worlds, manage their sustainability through TEK (Dove 2014). Anthropologists' understanding of people's knowledge of local weather patterns shows that it is more than an applied set of skills, enabling them to survive. Belief systems are also part of TEK. In small island communities, like Smith Island, TEK is a guiding principle for people's work activities, their migratory patterns, and their land use and management. Weather-related knowledge and a human desire to control weather are often expressed through seasonal practices, such as rainmaking rituals. These rituals are performed by a chosen person or a group of people believed to possess a gift which enables them to communicate with imagined supernatural powers. Smith Islanders' annual blessing ceremony in their church serves such a purpose. Community members gather in Tylerton's church for a blessing ceremony every year in April, just before crabbing season starts.

Tim Ingold (2000) notes that intuitive knowledge is not in conflict with scientific views. Instead, it derives, in specific environments, from perceptual skills that emerge for every being in the process of becoming. In his observations of Nordic hunters' knowledge of their environment, Ingold tells us that, in contrast to a formal, practical, authorized knowledge, hunters operate in "a sentient ecology" in relation with animals and other elements in the environment. Their knowledge is based on feelings, which arise from skills, sensitivities, and orientations that they have developed through their long experience of conducting life in a particular environment (Ingold 2000:25). Traditional knowledge is not an alternative scientific knowledge, but, rather, "a poetics of dwelling" (Ingold 2000:26). Smith Islanders' TEK includes their understanding of time in a place affected by uncertain, shifting weather and water patterns and a well-developed comprehension of the biodiversity in the Bay's ecology. The islanders' skill at weather reading is a climate-relevant example of TEK. Their safety and prosperity depend on their accurate weather reading of the sky. It is their ability to observe signs in the sky—for example, a lee set (blue clouds low on the horizon)—and then make an informed judgment about their movements in the seascape. Examining the different modalities of Smith Islanders' ecological knowledge and their methods of it passing on offers an insight into local perceptions of belonging and ways of being in the weather world.

A growing body of literature that explores place-based weather knowledge shows how, in many small rural communities, weather narratives are embodied in everyday weather experiences and probes that relational context (Harvey and Perry 2015; Paolisso 2003; Strauss and Orlove 2013). Collectively, these scholars emphasize localized weather knowledge that is expressed in place-based narratives: biographical memories of weather, folklore, weather talk, and storytelling. They argue that "weather is a function of place, and it follows that place has an important bearing on the weather and how weather variability and weather events are experienced and recalled" (Harvey and Perry 2015). "The weather is dynamic, always unfolding, ever changing in its currents, qualities of light and shade, and colors, alternately damp or dry, warm or cold, and so on" (Ingold 2011:73). In that same vein, Tim Ingold elaborates on Gregory Bateson's (1972) ideas of fluidity between the human mind and the materiality of the world humans inhabit. The weather, as Ingold put it, is not simply a powerful force to which humans and other living organisms and nonliving objects are exposed. Rather, it guides their ways of being as they strive for unity with the seascape rhythm, the "ecological unity" (G. Bateson 1972). Through stories and their evolving wisdom about weather, garnered from their experiential learning, islanders are passing along weather-related knowledge enmeshed in the climate and in habitat (Ingold 2011). To inhabit the weather world is to "mingle within" it: "In this world, the earth, far from providing a solid foundation for existence, appears to float like a fragile and ephemeral raft woven from the strands of terrestrial life, and suspended in the great sphere of the sky" (Ingold 2011:75). Weather can be predicted from *reading the sky*, but it can never be controlled. Therefore, multiple modalities of ecological knowledge—sensory, emotional, cosmological, and cognitive registers—guide Smith Islanders in their way of being in their weather world.

In the following chapters, I will explore the day-to-day lives of Smith Islanders and the extraordinary times they experience during their social events and rituals, all of which are experiences embodied in the ordinary. These are not universal experiences across time shared by multiple generations, but are lived or remembered ones during the time when I conducted my fieldwork (between 2013 and 2019). This book also includes a small number of additional observations and analytical comments drawn from my short-term visits between 2020 and 2023, but it is critical to recognize that Smith Island socioecology is part of an ever-evolving *local* process in relation to global processes.

ELMER "JUNIOR" EVANS

WORK IN THE SHANTY

2

Ways of Knowing

PASTOR JIM EVANS

On the wall in Jim's house hangs a framed photograph, capturing his father at work. He is doing what Smith Islanders call "fishing up," or tending to the live soft-shell crabs kept in the water in wooden floats. When I asked him about this photo, Jim walked over to it, took the picture off the wall, and held it in his hands, letting his memory take him back in time. He recalled specific times from his childhood when he was learning the traditional work on the water.

> This picture depicts the time when they used to shed the crabs directly in the water. It was a wonderful period of my time, how I learned being a waterman. What I knew, he taught me. Heritage of Smith Island means knowing how to be a waterman. I distinctly remember a culture of shedding the soft-shell crabs. It is what we lived upon, that was our livelihood. Being a waterman means knowing how to catch and shed the crabs, how to do it successfully.

This photograph is a part of his biography. His narratives, inspired by this image, bring him into the frontline of the story and show us the bond between his sense of self and his being in place. When Jim spoke about "the culture of shedding the soft-shell crabs," he referred not only to a specific traditional practice, but also to the knowledge—ecological, emotional, and sensory—that is specific to the Bay and Smith Island. All of this is defined by crabs, weather and climate patterns, water and land, smells and colors that are unique to the island, cyclical time, and economic systems. Jim talked further about the depth of his personal view of what it means to be a waterman: "Once you get it into you, being a waterman, it never really gets out of you. You are out on the sea, out on the water, just you and God. You have lots of time to look at nature and think about the harvest God provides."

Jim's testimony to his engagement with God when on the water and in the seascape is an excellent example of an internalized sense of self through faith, and of people, like Smith Islanders, who depend on their environment, which is always embodied in ecological systems. This is the concept of "way of being" that Jennings broadly defined in the previous chapter. Being a waterman is more than learning technical skills, as working in a seascape is dangerous. According to Jim, being successful means becoming socially connected with other watermen, in addition to having faith in God: "Being a waterman also means another thing. I remember all the captains, how watermen are there one for another. When one is in distress, the others come to help, that is something I remember, how close-knit community we are here. My father taught me to treat people good, I remember him telling me, 'Always be best you can be.'"

Knowing how to be a loyal member of the watermen's community is important, and it must be mastered by every individual if he is to sustain his livelihood in a small island community where social solidarity is an essential part of life. I could see a gentle smile on Jim's face and hear joyful softness in his voice when he remembered kinship stories rekindled by the family photographs. Yet when reading a photograph that reminded him of changes in the land and in life, a sudden sadness entered his voice. The loss of land through erosion is, to Jim, a sign of the possibility of losing his place.

> This picture means so much to me because it is all gone now, the land has receded, marshes are all gone. It is just something I will cherish. Because I can look at it, and just like a photographic memory I can see how it was that time, how we lived, how the community was. It was a booming place. It was an era of my life I can remember because [of] what I was taught as a youth. How to live as a human being, and how to treat other people.

While reading this old framed photograph hung on the wall in his living room, Jim told stories of his childhood, of evolving while learning to sustain himself in the seascape. The materiality of a photograph holds a social biography in which the notion of time overlaps the process of remembering. In the process of storying about becoming a Smith Islander, a waterman, and a human being, he told a tale about particular ways of being while belonging to the seascape and the community. "You have to be born on Smith Island to be a Smith Islander," residents say. This maxim expresses their awareness of the skills one must learn from relatives early in life if one is to do well on the island.

In this chapter I further explore the process of becoming. I place special emphasis on experiential learning, beginning in early childhood, and the sensory knowledge the islanders develop in relation to their environment. I take Smith Islanders' ways of knowing as a dynamic process embodied in their life course. I show how a changing socioecology has altered the potential for intergenerational relations, which older islanders cherish. I consider, too, how such changes in one's later years pose a new dilemma regarding aging for current Smith Islanders. I discuss the process of becoming a member of the Smith Island community through kinship and work skills. Presenting the life history narratives of a Smith Island waterman known as Eddie Boy, I explore how, in his view, experiential learning about work on the water and reliance on kinship have contributed to his developing a sense of self and place over his life course. In my analysis of local ecological and sensory knowledge, I show how labor in a kinship-based community defines the islanders' senses of self, which is integrated into their socioecological systems. In the second section, I discuss the kinship/aging dilemma that Eddie Boy shares with others on Smith Island.

Letting Her Go

Eddie Boy retired from working on the water in the summer of 2016. It was at the height of the summer season, on a hot and humid island morning, when I walked into his shanty. (A shanty is a working space, a wooden building dedicated to processing and storing a waterman's catch and doing other work, such as maintenance and recordkeeping.) To my surprise he was not there, so I stepped outside and looked for him on the boardwalk, where his boat was tied up. Eddie Boy was standing on the end of the pier by his shanty, watching two young men getting ready to leave in a boat. Having spent considerable time on the island by this point, I recognized that they were not from Smith Island. Eddie was intensely quiet, leaning on a piling while the boat was slowly pulling out of the slip. Having watched watermen on Smith Island operate their boats with practiced ease, I could tell that these strangers were just learning how to handle the craft. It was when the boat turned, and I could read the name *June* on the stern, that I realized what had just happened. Eddie Boy had sold his boat.

As the *June* headed out of the harbor, I followed Eddie Boy as he quietly walked inside the shanty. During early summer mornings, his wife June was always there with him, picking crabmeat. But that morning June was not there as Eddie Boy sat down in his chair, situated in the middle of the room. In his quietness, I realized that I had just witnessed an extraordinary event. I broke the silence by asking about the absence of his wife, and he said, "June went to the mainland. She's not happy about this, she's not ready, but I told her I am." He then continued, "I had to slow down first, but now it's time to stop. I had to let her [the boat] go." There was no regret in Eddie's voice, as he had come to accept that he had to sell his boat. Having witnessed the lifelong connections between a waterman's sense of self, his family, and his boat, I recognized the symbolic power of a boat in the men's lives. Eddie Boy's childhood memories shed a new light on what I had witnessed that summer morning when Eddie Boy made a step toward retirement and "let her go."

"I can't imagine living my life elsewhere," said Eddie Boy when I asked him about aging. "I take time to sit and observe nature and enjoy," he calmly remarked when he spoke about his sensory joy and personal satisfaction: "We live a simple life here, without major distractions and material conveniences people have on the mainland."

Emotional engagement with the landscape is not limited just to sources of food and shelter, but also includes sources of beauty, power, excitement, and other human values (Anderson 1996).

One evening, while sitting on a bench by his shanty at the end of the day, watching the sun set, Eddie Boy told me:

> Every year a group of men, well-to-do men, reserve a winter weekend in my shanty. They come from up north in February to Smith Island. They stay around the table here and shuck oysters while they keep the outside door open to watch the sun on the water. They come down here from their "worldly life," all the way to me, to seek something special. One of them told me that this is the best time in his whole year. For me this is what I have every day and all year around. I am happy. They come for one weekend. I have a year to feel my happiness.

When reflecting on his growing old in place, Eddie Boy often spoke of joy from gazing at the sky, watching the light on the water, and listening to the birds. He took time to sit and observe the island's landscape, a beauty he didn't take for granted. Such "mesmeric qualities" of water, as Veronica Strang (2004:51) put it, "are of particular interest in considering sensory perception and the creation of meaning." I could see how his stimulating day-to-day interactions with the seascape and the island's landscape resulted in a strong bond between Eddie Boy and his environment.

A year later, Eddie Boy was diagnosed with an advanced form of cancer. After his brief stay in a mainland nursing home, he decided to come back to his island house. Uncertain about how much time he had left to live, Eddie Boy remained inside his home, in the care of his wife and children, throughout the summer of 2017. For a while, Eddie Boy's sickness made him invisible to the public. The Smith Island community was distressed over his rapidly diminishing strength and his absence from public events. The community's

EDWARD "EDDIE BOY" EVANS

distress was intensified with the news that Eddie's brother, Junior, had also been diagnosed with cancer. Over his life course, Eddie Boy had established an important role in the Smith Island community. He was among the islanders who never left the island, even when others evacuated before a storm. Fellow islanders considered him and his brother Junior to be pillars of the community. "Eddie Boy is a very good waterman," they often said, and, emphasizing his wisdom, "one you can always ask for advice." During summer social events, Smith Islanders' collectively expressed compassion for the pain and physical suffering of the two brothers. It was apparent to me that their sickness affected the whole community.

The Male Life Course on Smith Island: Learning the Ways of the Waterman

For her size, she was about as smart a boat as there was on the Bay.

CAPT. EDWARD HARRISON

Edward, known as "Eddie Boy," was born on Smith Island in 1938. With the exception of two years spent in Baltimore during World War II, he has lived on Smith Island all his life. For Eddie Boy's generation, as for prior ones, survival on the island depended on absorbing traditional knowledge while growing up. He traces his ancestors back eight generations, to the eastern shore of Virginia on his mother's side, and to the western shore of Virginia on his father's side. Eddie Boy recounted how he first recognized his desire to become a waterman.

> Well, all my life from when I was a small kid, I always desired to grow up, naturally, like most children do, to be like their father. I mean my father and grandfather was with me like anybody else. They were role models with me. I never cared to go to school if I could get out and work on the water. I always had a desire from a very early age to have my own boat and just be my own boss. Now, at that time you didn't look at it as being your own boss; it was more or less the freedom part of it.

From the perspective of his childhood memories, Eddie Boy recalled his first taste of being on the water as a feeling of power and freedom. But learning how to captain a boat is not just developing certain skills, it is part of the larger process of ways of knowing, which include acquiring social, sensory, and emotional knowledge, as well as learning a trade and becoming familiar with the environment (see Berkes 2018). Eddie's admiration for his father and grandfather is not exceptional among Smith Islanders. As I noticed during my fieldwork, other watermen vividly recalled their childhood memories of togetherness and of closeness with their kinsmen during the learning process. For most, it is their strong relationship with their fathers, other male kinsmen, or mentors, such as a boat captain. A boat is an essential part of a man's life course on Smith Island. Eddie Boy reflected on the process of developing ways of knowing through learning how to captain a boat, a critical skill for Smith Island boys entering manhood.

> And pretty well every kid around here, when they got [to] 8, 10 years old, in between there, had a skiff of some type. Now, some was fortunate enough to have new ones, some was lucky enough for to be able to find one that somebody had throwed away and they'd fix it up and whatever. But we were always just messing around with these type of skiffs, riding 'em like our fathers' boats and what have you, and it was a lot of fun, but it also built the desire into you to be competitive, because you always wanted a boat that was better than the next boy's that was your age, and you were trying to do something different than he was or trying to go faster and things of that nature.
>
> I would say that when I was no more than 9 years old, I had my first skiff, which was a big thing for us at that age. It's more or less like a kid today when you get 16 and you get your first automobile or something. It was really a big deal with us when we got our first skiff. Probably didn't cost over 30 or 40 dollars. Didn't have a motor on it, but at that age we could take that skiff here in the shallows all around the island with a sail on it. We were always rigging sail on it, and it was something that your parents pretty well let you do your own thing. If you wanted to put a funny-looking sail on it that was fine. It was your boat.

Eddie Boy's recollections of learning to operate a small boat, a skiff, provide an insight into the process of becoming a Smith Island waterman. When Eddie Boy was growing up, getting that first skiff was a rite of passage on Smith Island. It meant learning traditional skills, as well as gaining sensory knowledge and emotional satisfaction from a child's desire to be like his parents. It was a first taste of being empowered and feeling "freedom," as Eddie Boy put it. It was being away from parental restrictions while learning "the seascape's rhythm" (Maurstad 2004) and sensing in various ways in which one's body responds to the environment. "Knowledge of places is therefore closely linked to knowledge of the self, to grasping one's own community and to securing a confident sense of who one is as a person," writes Keith Basso (1996:34) in a different ethnographic context.

Becoming a Smith Islander and a waterman is a process of experiential learning that stretches over the course of childhood and gradually extends into manhood. Buddy, one of the middle-aged watermen, remembered how his dad would take him to work on his boat: "My dad put me in the box on the boat and I would be with him all day. I even fell asleep and was mad when my sister woke me up when we arrived home in the afternoon. I didn't want them to know that I was sleeping, but it was all so good." Mark, one of Buddy's contemporaries, remembered that "as kids we would play in a skiff catching crabs with nets. We would also go dock to dock when we didn't have a skiff, catching crabs hiding near pilings. Later when I began with my father at first, I would do small jobs like steering the boat, throwing buoys overboard, whatever was helpful to my dad and the mate he had working with him."

Learning the rhythm and skills of their craft, starting from an early stage and extending over one's life course, is a pattern in many of the childhood memories I collected from Smith Island's three communities. Those Smith Island men with whom I had the opportunity to talk remembered their childhood play as related to the work their fathers did and their initiation into the process of becoming a waterman. Some men discontinued their seasonal crabbing and oystering over their life course, yet their childhood narratives reveal similar patterns of learning. Bobby, a tugboat captain, described how his dad would teach him to control his skiff: "One day Dad asked me to take crabs to the shanty. Our shanty was out on the water near marshes on your way to Rhodes Point. As I was coming closer, I stopped the engine and let her drift slowly toward the shanty. 'You have to get her there much faster than that,' he let me know."

For watermen, experiential knowledge of the seascape is part of the traditional knowledge they learn from older male kin. Smith Island watermen spend most of their lives on the water, and it is their extensive sensory knowledge of the seascape that shapes their exclusively male ways of knowing. When defining their work on the water, Eddie Boy and others relied on their sensory experiences, which are drawn from their closeness to nature, and to water in particular. Mark, another waterman, indicated that there are certain smells around the island at different times that he can relate to: "Honeysuckle smell in June while working near the island, near Ewell, with other blooming bushes in May and June reminds me of the end of school year, when I was a kid." For both men and women, sensory experiences on the island—expressed as the fabric of everyday life in its sounds, smells, tastes, and sights—are all shared elements of a Smith Islander's sense of place. The earthy smell of marshland; the scent of freshly washed clothes and perfumes in church; the taste of local food; the sounds of the wind crossing grasslands, water pushed onto the island's shore by the wind or trickling over swimming crabs in shanties, and the sputter of crab boat engines starting out before sunrise and returning from work in the heat of midday, these are all parts of "sensual cultural complexities" (Stoller 1989).

Eddie Boy's memories show how early experiences with a skiff lead to a quick transition from mimetic and ritualized play on the water to its end stage: a rite of passage to an assigned job. This is how he recalled his first work trip to the mainland.

> The first time that I had—and I remember it real distinctly—I was 12 years old, and my father let me go to Crisfield the first time by myself in a big boat to take the crabs [a one-hour boat trip]. And looking back on it, I can see where it's pretty young, you know, to trust a large boat to go to Crisfield, even though I'm quite sure my father had thought this out plainly, and he picked the type of day that he says it's okay to do, that he felt that it was going to be a beautiful day and you could see a long ways and them type of things. But still, looking back, it was pretty young for to be trusted with that. I didn't trust mine with my boat at that age. I mean, they were a little bit older than that before I would trust 'em with that type of a boat. But anyhow, my father did.

We can see here how this first work assignment quickly turns from childhood play on the water to the adult experience of taking responsibility. We can also see the kind of knowledge an older generation of Smith Islanders must learn early on in their lives. Eddie Boy gives us a suggestion here: while mastering a boat on the water is a critical skill for watermen, acquiring sensory knowledge is also about embracing and overcoming an ever-present solitude and a fear of death when working on the water. In looking back on his youth and his father's approach, but describing this now from the position of a parent, Eddie offered some further reflections.

> He always gave me the freedom to do these types of things, and it was a good thing that he did, because by the time I was 16 years old, my father became seriously ill and he couldn't work for two years. I had to take charge of the family at 16 years old. My brother was 14 and I was 16, and we'd take that boat, I was in charge because I was the oldest. My father made that very clear. And we managed, believe or not, to look out for our

family, and we had a large family. It was six children of us and plus mom and dad, and dad was sick. And we managed to keep that family together with enough—put food on the table and all the other expenses going for them two years. So if we look back, then my father was wise to let me learn at an early age. . . . Naturally, after that, as I began to get a little older, then I wanted my own boat. And at 18 years old, I had my own boat, my first boat.

Eddie Boy's life on the water and on the island continued with his marriage to June and the birth of their five children. By reflecting on his memories, Eddie Boy has shown how, on Smith Island, men's relationships with their kin and their boats are formed in early childhood and continue to develop well into manhood. His brother Junior later added:

"When my brother [Eddie Boy] and I would work on our father's boat, he was the captain, and we each had our own role. It was five of us. I was a mechanic in charge of engine. One day, engine would not start, and so I cursed out of my frustration. I never forget as long I live how my dad put his head down, and he didn't say nothing, but I was so ashamed for it, and I never used that word again. My father knew how to handle situation very well and I learned from him and would do it just like it with my son."

When I first met Eddy Boy, he was in his seventies, a working waterman and a respected leading elder in his community. His narratives led me to further explore others' memories of childhood that reflected experiential learning and the role of a boat in the life course of a Smith Islander. I found that, during childhood, acquiring skill by playing with a skiff was preceded by much earlier play with a small toy boat. The toy boats were carved from one piece of wood and then painted, and boys would pull them with a rope.

Jennifer, Eddie Boy's sister, remembered when her son got his first tall boots, which made it possible for him to play in the backyard at high tide: "He was there for hours, playing in the water with his tall boots on. After he set up the stations in line and connected them with a rope, he pushed his little toy boat from station to station, playing crab pots pulling." What makes Jennifer's story such an interesting example about childhood on Smith Island is the interconnection between imaginative play and experiential learning by observing crabbing skills.

For some, a link with the actual workboats is also manifested in their interest in workboat models. When visiting Smith Island homes, I noticed these models displayed in their living rooms. Some of the models are passed down from the previous generation, while others are more recent. They are considered to be a form of art, a cultural heritage on display during local art festivals and events. Skillfully crafting workboat models is a creative activity many islanders share with their communities, as well as with their relatives and friends on the mainland.

Because of their location and livelihood, as Eddie Boy put it, "a boat has a very significant place in the life of a Smith Islander." Imaginative play in early childhood mimics adulthood. The ownership of a small skiff provides children with their first taste of freedom, competition, adventure, and desired adulthood. Work on the water begins during childhood and continues into one's later years, and a boat is one of the central elements of family sustainability. Smith Island workboats are cared for daily. Watermen clean their boats every day and do small-scale repairs during the crabbing and oystering season; large-scale maintenance of the boats is done during the offseason.

In addition to their care for each workboat and their pride in the model boats that are symbolic of their cultural heritage, I noticed the watermen's emotional connections to them. Watermen refer to a boat as "she" and name their boats after their wives, mothers, or daughters. Smith Island's personalized workboats emphasize how each boat has an agency, representing its anima. Glenn Lawson, a local who has written a book about Smith Island, notes that a boat was always treated like an important member of the family and handled with respect: "Being on the boat all your life, that boat becomes part of you. Helps raise your family" (quoted in Johnson 1992:12).

Jerry reminded me of this close connection between boat and family (or individual person) when he showed me an old snapshot of his first boat: "This was my first working boat. I named her the *Morning Glory*, after the flowers. There was a lot of them growing around my shanty and they would be always open in the morning. Last year, it was the first time that she [the boat] sunk, in my whole life." He paused, and then said, "I was devastated, a bad wind come up." A slightly faded snapshot of Jerry on his first boat brought a pleasant smile to his face. It was a reminder of his "initiation to a man's adulthood," as Eddie Boy put it when he talked about his life course and his boat.

While collecting old photographs, I found a photo of a little boy alone on a small boat and imagined how, for an islander, this image evoked a sense of pride and accomplishment. It was part of this boy's experiential learning. For islanders, being in contact with water, in addition to mastering the skills of how to manage a boat, is a vital form of day-to-day, close sensory interaction. Some Smith Islanders told me that their attitude of "going with the flow" is their way of responding to the movements of the water in their seascape. These sensory skills are inherently necessary for adaptation to the island's environment. People's analogies about movement, both in life on the water and on the island, are accounts of sensory ways of knowing, reflecting their outlook on life. For Smith Islanders, water is a powerful metaphor that permeates their ideas, emotions, and imaginations. Insights regarding water flow beyond its physical properties. For Smith Islanders, water is a treat, as well as sustenance for their way of life.

Watermen often talk about feelings of freedom on the water. The notion of freedom, embodied in a man's boat and his work on the water, is one of the significant patterns in Eddie Boy's life-course narratives, and it is a feeling is shared by his fellow watermen. Dwight, Eddie Boy's brother-in-law, spoke of how a sense of freedom comes with living and working on the water: "When you look from your window, here on the island, you see what is happening outside on the water, and right then you can tell how weather will affect everything. You are free to go and do what you need to do. You have knowledge and experience and act on it as you please with flexibility." Although people originally settled on Smith Island with the intent to farm, Chris, a former waterman, said: "Smith Islanders are not farmers, they have heart of the hunter."

The feeling of freedom the islanders experience, combined with their practical knowledge of ecological diversity and change, and their sensory knowledge of the seascape, drive the islanders' own flexibility to interact with their environment. For Eddie Boy, Dwight, and others, this sense of freedom derives from their ability to move physically with the water and weather, as they are a part of this ecology. It is what Jennings alluded to in some of his conversations, and what Anita Maurstad (2004) defines so well as temporal rhythms and belonging to a particular lifeway.

As it is in many diverse coastal communities, traditional knowledge on Smith Island is an evolving process, one which is learned through doing—that is, the experiential learning I emphasize in multiple chapters of this book. A cumulative body of knowledge, based on collective experiences and observations that are brought together, interpreted, and shared, produces what Fikret Berkes (2015:229) defines as community knowledge. The opportunity to pass on their knowledge to the next generation is something that is highly valued by Smith Island watermen and seems to be a significant aspect of their work. They are well known for it in the Chesapeake Bay community (Warner 1976). From Eddie Boy's and others' narratives, we can see how, from early boyhood to late in life, work shapes the life-course obligations in the life of a Smith Island resident.

Growing Old as a Waterman

Like other aging watermen, Eddie Boy maintained his excitement about work into his late seventies. While admitting that his physical limitations frustrated him, he tried to retain the status of a working waterman into his later years. Because work and individual independence are generally highly valued on Smith Island, the main occupations for older adults become those of maintaining economic independence, sustaining their heritage, and managing one's work and health situations. On the island, some watermen meet for early morning coffee in the local shop and talk while they listen to the marine radio and television. In the company of others, they enjoy telling stories and jokes, discussing their work and the weather, and supporting each other in overcoming hardships.

The migration of younger generations to the mainland has had a profound influence on their aging parents and grandparents, who have remained on the island. Today, elderly Smith Islanders are still active storytellers in the community, but in the absence of their children and grandchildren, they are losing the opportunity to pass their stories on to the next generation. While older adults probably cannot imagine spending their lives in any other place, engaging in different work, or losing their Smith Island community, they have supported their children in seeking education and finding new professions on the mainland. Most of Eddie Boy's children's generation left the island for work on the mainland. His grandchildren are not pursuing careers on the water, but instead are seeking higher education. They visit the island sporadically, usually only one or two times per year. Typically, their parents and grandparents travel to the mainland to participate in their grandchildren's lives or for holidays. In any kinship-based community, separation of the family by migration presents a dilemma, which is even more acute for the super-aging Smith Island community (Danely and Lynch 2013).

Eddie Boy and his wife June have been proud of their kids. Eddie Boy said multiple times that "my granddaughter is doing really well at the University of Baltimore Law School. She already has an internship and her article was accepted to a professional journal; she is doing well! As far as I know she is the first lawyer from Smith Island, and when the judge picked her for an internship, he told her that a reason he picked her is because he knew she must have worked twice as much than others to get where she is." But in watching their grandchildren succeed on the mainland, Eddie Boy and June are caught in a kinship/community dilemma. For Eddie Boy, like others on Smith Island, witnessing his children and grandchildren leaving the island, first for school and then for jobs, foretells the absence of people on Smith Island in the future. While excited about his granddaughter's success, he also knows that there is a painful price to pay: the likely discontinuation of his community. Struggling to hold on to their stories in the "absence of the listener is like the deprivation of an individual. It is as if all creation stories are stories of separation" (Myerhoff 2007:20).

Smith Island storytelling is an integral part of active agency for older people. Caitrin Lynch (2012:199) points out that agency and control are two nouns that may not be typically associated in American society with older adults: "Whereas they are important values in the United States, Americans seem to assume stereotypically that old people do not have agency: their bodies are failing, they are not in control, their life choices have already been made, and they are just finishing up their years." Smith Islanders' engagement in active work provides them with a sense of purpose, "an agency in their late adulthood" (Danely and Lynch 2013) and contributes to their "active and energetic later life" (M. Bateson 2013). Yet their obligation to carry on the island's heritage is now thwarted by their lack of opportunity to pass old stories on to the next generation. It leaves them with dilemmas: personally, in feeling unfulfilled, and collectively, with a sense of losing their traditional means of reinventing their community.

Eddie Boy died in January 2018. At his funeral, silence overpowered the traditional storytelling. At a Smith Island funeral, people typically share stories and even tell jokes. At Eddie Boy's funeral, however, all the designated speakers struggled with the sound of their voices. It was as if all speech was silenced by the pain of losing him. I recall only one message, which was like a collective echo across the crowded church in Ewell: "He was a person to go to, always available to share his wisdom." People agree that Eddie Boy's accessibility and ability to share his knowledge with others has become, in the eyes of Smith Islanders, a heroic way of being. His skill at surviving made him a successful waterman, yet at his funeral, I realized that the triumph of his life was his service to his community, the place where he belonged. Eddie Boy's life history is an account of his personal experiences over his life course in relation to a place defined by a unique socioecology.

My focus on Smith Islanders' life histories provides a larger context for a better understanding of how traditional knowledge is constituted in the life course of individuals connected to a place. Reading Eddie Boy's life history and tracking narrative speech patterns about his work on the water opened a new space for my further exploration of traditional, ecological, and sensory knowledge, as well as, in a broad sense, a symbolic embodiment of boats in the life course of Smith Island males. By examining Eddie Boy's memories of his childhood, I show, from the perspective of his life course, how the process of his becoming an adult waterman is defined by intergenerational practices that are structured by knowledge. I also demonstrate how knowledge, embodied in the life course of an older person, Eddie Boy, underwent major shifts related to socioecological changes.

> We are, in a sense, the place-worlds we imagine.
> KEITH BASSO, *Wisdom Sits in Places*

Smith Islanders' feelings of belonging to the island give them a deep sense of being connected to their ancestors, their cultural traditions, their work, and their community. To continue their traditional island life into late adulthood, they must remain active, able to manage both their work and their health. They must do all this while bearing witness to the erosion of their land and the Smith Island community. As his case illustrates, Eddie Boy's emotional engagement with his boat, the seascape, and the island were essential parts of his means of successfully aging in place. Yet, as I see it, trends toward change also presented him with a kinship dilemma, related to the discontinuity of his own heritage. These are elements that shape life-course experiences. Eddie Boy's and others' narratives provide insight into the deep relationship between self and place, although it may not apply to all Smith Islanders. Nonetheless, we are left with a deeper understanding of how islanders belong to their ecology, the critical role of kinship in their sustainability, and the complex set of dilemmas faced by the aging population on Smith Island. The island presents a closer look at difficult feelings in an aging community, where elders, in the absence of a future generation, are still holding on to their stories and

HARBOR AND SHANTY

traditions. They are facing not only their own physical death, but also the death of their island. The stories, told here by Jim in the context of old family photographs, and by Eddie Boy in recounting his life history, demonstrate how their sense of self is strongly connected to their sense of place. "Knowledge of places is closely linked to knowledge of the self, to grasping one's own community and to securing a confident sense of who one is as a person," writes Keith Basso (1996:34), although in a different ethnographic context, about notions of self and place. All the stories remembered from childhood, which shaped Jim's sense of who he is today, are bound up in the vulnerable, unstable present state of environmental change. In this light, the erosion of Smith Island is not only a threat to the land, but also to the islanders' sense of their place within the world. Growing old in one's place, in light of rapid socioecological transformation, is raising a new question about the future hope of this and other small island communities.

3

Land and Water

WATERFRONT. *Source*: Somerset County Library

Reading Landlines

"To us, the island is such a small and personal space. You become familiar with every detail of the entire landscape," said Bobby when looking at the photographs in his grandmother's albums. He pointed to one photo showing his grandmother standing by the waterfront, explaining that "'keeping up your waterfront' was something people would say in reference to how well you take care of your yardland." During our conversation about the island's grounds, he reached for his cellphone and pointed to visible lines a person can follow from point to point on a satellite map of Smith Island: "These are old ditches defining people's properties. You can see all the places where one-time people used to live. The houses were more scattered in this area in the past. The lines on the map show where the ditches are, and these were manmade to elevate the land and protect the houses." When speaking about the photographs, Bobby, like other islanders, noticed the smallest details about landscape changes on the island. They indicate the movement of the ocean's water, which is eroding Smith Island's grounds. In this sense, albums and photos open a new space for telling the story of changes in the land, which otherwise is kept in silence. The "figure of the island" (Pugh and David Chandler 2021)—which these authors refer to as "liminal and transgressive space"—points to other islands worldwide, suggesting that an island is a key site in enabling alternative conceptual forms for what has existed in the Anthropocene when thinking about climatic futures for people in blue places (Foley et al. 2019).

"Place is a way of understanding the web of interrelationships between humans and landscape that shapes both humans and the landscape through time," writes Sarah King (2014). In this chapter, I explore the multiple ways in which Smith Islanders trace the landscape and manage their land. Reflecting on my conversations with Bobby and other islanders, I argue that their sense of self is embodied in the multitude of their engagements with the materiality of Smith Island's grounds, which, to islanders, holds sensory experiences. I follow stories, such as those attached to fruit trees in remembered places, by examining childhood memories that are connected to sensory or moral recollections that are embodied in places, in order to reveal lived subjectivity in this island community. I discuss notions of self and the fertile sustainability of the land over time when I introduce conversations with Eddie Boy and his brother Junior, Char and Jerry's narratives, Everett's plans, and Carol Ann's introspections. I analyze how the islanders' sensory knowledge and memories are integrated into their soundscape. Exploring sensory experiences in "a zone between earth and sky where all happens" (Ingold 2011), I connect subjective poetics embodied in the materiality of Smith Island's weather world. The story about Everett's gardening project, entitled "Small Beginnings," follows his journey in promising the recovery of sustainable gardening on Smith Island, which has gradually been lost over time to increasing periods of flooding.

Storied Grounds

Jerry reached into his box of old family photographs and held up one of the faded color snapshots from his childhood. His eyes wandered across the image of him as a little boy, standing in the yard and holding a toy in his hands.

JERRY SMITH

Land and Water

> You see, behind me there is a rosebush. You can never have anything like that growing now. When I was a kid we had a pear tree in our yard, we also grew corn and tomatoes. You would not even think about doing that now. The tide would ruin it. The land level is low and ocean level higher than before. Until the '80s, we would have high tides in the early spring and later fall tides, and in between you could grow corn and tomatoes. You didn't have to worry about wind from the east pushing tide into your yard.

Jerry continued to reminisce about the seasonal change in water patterns and its impact on land, in addition to telling stories recounting memories about his family. When I first saw his family photographs, I didn't anticipate that Jerry's small color snapshots would elicit stories of the land and changing patterns of water lines. I expected childhood stories coded within the old family frames. Such stories surfaced in most conversations, but Jerry and, later, many

JERRY SMITH AS A CHILD

others also revealed how their personal experiences are attached to places on the island. I could see how their sense of self is intertwined with the physical sites. Stories about the island, constituted in memory, are framed within material objects, such as photographs.

One time when I visited Charlotte, known on the island as "Char," she was holding an old framed photograph of her mother. She took me back in time to the geography of her childhood as she said with a smile, "In some of my memories the island was a bigger place than it seems today. My mother would not let me go to 'down the field' [a name for the southeastern part of Ewell]. To her that was a dangerous world for a young girl. Down the field was perceived as too far and unknown." Smith Island women have different relationships with the island's grounds than men. While the men work daily on the water and identify with the movement of the sea, the women trace the island's roads as they walk or drive on the land they identify with. There are a few exceptions, such as when women will sometimes need to pilot a small boat between villages separated by a water canal. They may even fish up crab flats. Both of these activities require women to master the skills necessary to operate a small skiff, and girls learned about the seascape from playing with skiffs with their brothers, as they are not expected to work on the water.

In general, it is the women who care for the land, and for some girls, the process of becoming an island woman is defined by cutting grass and tending to the yards. Carol Ann, for example, talked about her first job, cutting grass, which she considered to be a significant point in her process of becoming an adult. Tending to the grass saves the land from marshland overgrowth and faster erosion, and it reduces the breeding areas of flies and mosquitoes. Therefore, well-maintained yards are an essential part of local knowledge about care for the environment and public health on the island. For a female who is coming of age, the significance of keeping grass short on Smith Island is less of an aesthetic concern and more a means of learning to care for the environment. Bobby's, Jerry's, and Char's readings of the old photographs are more than a forensic exercise. Their pictorial narratives show how land management and land care are intertwined with the emotions they experienced at different stages of their lives. When we look beyond a snapshot of an extraordinary moment and notice many details in the landscapes that are captured in family photographs, Jerry's and Bobby's narratives present an ethnographically rich point, one that offers insight into their storying of the land in the weather world they inherited.

In the spring of 2020, I stopped by to visit Jerry. While standing in his front yard, he returned to our conversation about the island's changing environment: "You see, my front yard is lower now, it wasn't like that when I was a kid, we had many trees. If I left my yard alone [now], it would turn into marsh. If you don't cut the grass, it would grow the water bushes." He walked closer to his house and pointed out that, "when you look at this stone, you can see how deep the stone sunk into the ground." Jerry again brought up how the land has changed since his childhood and proved his point by referring to his earlier narratives, related to the childhood photographs from his family album. He openly talked to me about climate change and its impact on the island, while others may remain silent or openly declare their disbelief. But it is important to note that Smith Islanders do not hold homogenous perspectives and concepts related to environmental issues and climate change. When Smith Islanders turn their albums' pages, visual and narrative patterns emerge and intersect. Their family photographs hold a double register of emotions: joy from reading social biographies in their kinship lines, and expressions of pain elicited by the images of their intimate landscapes dissolving into the sea through relentless erosion.

As I sought to understand the islanders' forceful, deep concern to sustain their island life, Bobby's reading of the lines on the satellite map of Smith Island led me to think further about new questions related to the residents' engagement with their land. I was interested in learning more about how they interacted with the materiality of their land. Old drainage ditches have lost their original purpose, as the dwellings they supported were condemned, due to their instability, which was caused by severe erosion. But in their visibility on the current maps, these lines incised into the landscape reveal a lost thread in the island's settlement patterns. By outlining former properties that have now vanished, these lines tell the story of the islanders' land management and their movement across the Smith Island landscape. If the lines made by people are more than just human traces, but instead are an available form of material culture, as Tim Ingold (2007) contends, I further argue that reading inscribed landlines in so-called ditch banks from satellite maps offers an important insight into place-specific land management and resettlement movements. Because Smith Island is essentially at sea level, the smallest amount of rain contributes to higher water accumulation in some parts of the island, further reducing the landmass of the island. Beyond the archeology of historical dwelling patterns, reading landlines teaches us about the place-specific practices islanders developed as they were—and still are—challenged by the constant shifts in their island's ground.

Historically, land on Smith Island has been managed by individual families, who have cared for the properties surrounding their houses, gardens, or small farms. The community-built channels across the island ease connections between settlements. Yet growing problems with rapid erosion have resulted in the islanders' need to take new approaches to land management. Erosion on the northwestern side of the island and high-water levels on the eastern side have become critical focal points in the Smith Islanders' efforts to improve the resilience of their land. Erosion poses a serious threat to them, but because they have historically managed their land, they approach this current issue as a problem they will be able to solve.

In recognition of the islanders' persistence in land management, the US government funded the construction of a seawall by the Army Corps of Engineers in two areas on the northwestern side of the island. Additional funding was approved for an environmental engineering company to investigate a land-erosion improvement plan for the entire island.

In addition to federally funded stone walls protecting the island from the northwest, Smith Islanders employ multiple short-term strategies to address high-tide issues affecting both of the island's public roads. For example, when I visited in the fall of 2019 and the spring of 2020, I noticed that the residents had made new road ditches in several places on the island in their efforts to control water accumulation from high tides and rain. I saw a man in a small truck moving bags of sand from the North End and distributing the sand to areas on the island needing a higher elevation. Building up the strategic resilience of the land, as is evident from hidden old ditches lines traditionally established around each house during the island's early settlement history, is part of Smith Islanders' inherited traditional knowledge. Following human traces in the landscape through these landlines, which are a form of material culture incised into the remaining eroding grounds, we can see the islanders' ways of managing the land across time and place, as well as the limits posed by shifting land.

Searching for Sustainable Land: "Small Beginnings"

When Everett and his family returned to Smith Island after his years serving as pastor for a mainland church, he started a small garden next to the parsonage. He and his wife Carol Ann have a Facebook page called "Small Beginnings." During our conversation, I asked about the meaning of this title. He explained, "One time in our old parish on the mainland, a woman came to our door and said, 'Don't despise a small beginning.' She spoke about the verse from the Bible, from Zachariah chapter 4, verse 10. She was asking us to consider taking a small step in what you/we can do. So, to us, Carol Ann's health, our family's well-being, and gardening, all has been about taking small steps." Carol Ann posts daily pictures of their family meals, often featuring produce from their garden and other homemade ingredients. Gardening has become central to their commitment to stay healthy by eating natural, unprocessed food.

Everett, originally from Rhodes Point, transitioned from working as a waterman to pastoring for the Methodist Church. When he married Carol Ann, they first lived in her hometown of Tylerton. Everett's pastoring posts took them to Tilghman Island, a fishing community similar to Smith Island, in Chesapeake Bay. By the time Everett and Carol Ann had moved back to Smith Island in 2017, cultivating their own garden and eating the food they harvested had already become a habit: "I remember when we lived in Tylerton, my wife would grow flowers. She would bring them home. Well, I asked her if she could bring me something I can eat. One day she said, 'Did you see my flowering cabbages? But you can't eat it, they are decorative,' she told me. She was after beauty."

Everett had really started gardening after he was assigned to another parish off Smith Island. "My wife was really sick, and we had to address her health by changing what we eat," he said, describing his motivation to begin growing vegetables. "I asked my family what they want me to grow. If there is one vegetable I will grow, what it would be? They all picked brussels sprouts. What I didn't know was that it is the hardest vegetable to grow, and I picked a bad spot. . . . I got really angry with myself, thinking, if everybody can drop a seed and grow something, why not me?" Everett also recounted the turning point in his gardening efforts. After four years, his efforts turned into what he called a "real garden," and when the time came to move back to Smith Island, he was confident about planting seeds and growing vegetables: "Now that we are here, I remember what my dad told me about his childhood and how it used to be on Tangier Island, where he is from. His family always had their own food from the garden and when others were in need, his mother would share and give to others." Gardening reminds Everett of what he had heard about the old days, and he finds excitement in being able to return to what his family did long ago to survive. But the real reason for gardening, Everett emphasized, is that he can provide organic food for his family. Two years after returning to Smith Island, he has been expanding his efforts by raising chickens, building a smoker to try to smoke fish, and planning to raise rabbits in the future: "I love the challenge and the creativity of it. Most everything in the yard is from an item I found. A chicken coop and fence, compost bins, and a fenced-in garden area. The creativity also comes with making the planting beds and preparing them with proper nutrients."

He then added, "I used to work with my hands. Crabbing is hard work. Now that I think of it, gardening is like crabbing. The proper preparation needs to take place in order to have success, having the right equipment and resources, putting in work and reaping a harvest." For Everett, though, gardening is also a spiritual endeavor: "I am still amazed at how I can plant one tomato seed, harvest tomatoes from that one plant to eat all throughout the summer, and save the seeds from all those tomatoes. From one seed comes food for my family and hundreds of seeds to replant the following year and give some to family and friends. Growing from seeds helps me see the blessings of God." While seeking nutritious food for himself and his family, Everett also came to realize how gardening offers him a spiritual experience: connecting a successful harvest with God's blessing of seeds.

Everett's reflections on his gardening experiences bring us back to the question of knowledge and faith I explored through Jennings's and others' narratives in chapter 1. Following Everett's description of "small steps" in his gardening, we can see how, through his engagement with the land, his knowledge, emotions, and spirituality mutually evolved. They even resemble the crabbing he did in a different stage of his life. Gardening practices seem lost in the face of the land's subsequent erosion, and the cultivation of fresh food has been replaced by grocery shopping at stores on the mainland. This correlates with a national trend toward a centralized food system and is not exclusive to Smith Island, but it is certainly exacerbated by erosion and the sinking ground level on the island. Everett's gardening efforts, driven by his desire for unprocessed food, have raised the possibility of recovering the previous tradition of gardening on the island. Although most Smith Islanders still make traditional foods for their celebrations and special events, fast food has become part of their day-to-day dietary routine. Everett's hope for sustainable food practices, materialized through his return to land cultivation as part of his search for healthy food, leaves us with further questions about the process of reinventing heritage, a question I will explore further in the following chapters.

Over the past decade, a small number of Smith Islanders have practiced gardening, planting seeds in elevated dirt beds to protect their gardens from flooding. Across the island, people's gardens are affected in different ways by rising water. In some of their yards, erosion has drastically changed the conditions of the land, but in other areas this is less evident. Overall, though, most of the island has lost its fertility. People continue to manage their land by cutting the grass and maintaining their waterfronts, some still harvest fruit from their fig and pomegranate trees and cultivate a small number of flowers, and still others return to their secret places on the island to harvest wild asparagus in the spring.

The Ghost Trees and the Soundscape

In the places where marshland frequently penetrates the island, the line between water and land is shifting. As a result, the land is eroding. Island trees lose nourishment and are slowly dying from this erosion, which affects all island inhabitants. For them, dead tree trunks sticking up from the ground, known by ecologists as *ghost trees*, are painful symbols of the island's changing ecology. The ghost trees not only are powerful symbols of the land's lost fertility, but they also have changed the soundscape on the island.

GHOST TREE

In the spring of 2020, I visited Jerry in Rhodes Point. Sitting on a bench in his front yard, during our conversation I noticed the silence of the soundscape and remarked, "In Ewell [the village where I stay on the island], the birds have been singing a lot." Jerry quickly responded, "That is because we don't have many trees here. Healthy big trees attract birds to gather in the spring. I hate to say it, but I watched when they built houses. They took down trees here in Rhodes Point, and then flooding contributed to the death of the rest of our trees."

I was listening to a quiet breeze blowing in from the water, and I thought about the sounds that were lost with the loss of trees on the island. It is through the experiences of our senses that I and other newcomers can relate best to the Smith Islanders' sensory knowledge. When I take a walk across the island, inhaling the earthy smells from the marshland, I also listen to the soundscape. I always experience the intensity of the wind and the susurrations tall marshland grasses make when bending through the air and reaching closer to the ground. The sounds of seagulls in the morning wind, and of engines on the workboats, going out at dawn and coming back from the sea in the afternoon, all echo in the wind. And, while the joy of being there has a different meaning for islanders than for visitors like me, these are pleasant sensory experiences, now also familiar to me, that I share with the islanders when I stay on the island. Seeing the lines in the seascape that are drawn by the light is visually impressive at any time or season.

I recall a conversation with Eddie Boy when he talked about aging and the joy he found in listening to the birds coming into his garden in Ewell. It was an early summer evening in 2017, just before sunset, when waterman Eddie Boy and I sat on the bench by his shanty. The thought of fruit trees, now lost, evoked special memories of his favorite taste sensations. As Eddie Boy remembered, "At the North End, there used to be a big farm and an orchard with different fruit trees. One of my best memories is the taste of plum jam with heavy cream on top. That was my favorite." A few months later, when I came to visit Eddie Boy one morning, he shared another story related to the taste of things: "There is an old plum tree in Tylerton. My sister Mary Aida sent me a basket of plums, and June [his wife] made jam for me. I was sitting alone and slowly eating the plum jam on toast. The taste took me to memories far back in time. I remember games we played; nobody knows [them] in this generation." Like Eddie Boy, older islanders still remember the tastes of the food this old farm provided when they were growing up. When I was recording these memories, I found that not only are some tastes specifically associated with a place in time, but they also point to the relationship between subjectivity and places. Junior, Eddie Boy's younger brother, recalled one such instance.

> When I was little, everybody grew their fruit, and there was a woman on the island growing white grapes in her garden. Other people had grapes, but she was the only one that had this type, the white grapes. I went with other kids to her garden and while stealing her grapes, she caught us. I remember all my life what she told us. "You can eat as much you wish, but don't waste any!" she said. I always remembered how she handled us kids stealing.

Junior's memory of a woman's generosity, manifested in the sweetness of a gift of white grapes, was transmuted into a lifetime of sensory knowledge. Junior recalled another, similar story: "We use to make shortcuts across the land when we were kids, and there was always somebody who didn't like it. One time, we kids crossed the land of a man. The guy was sitting under the tree in his yard and saw us, and so I asked him if we can cross his land. He said something I would never forget. 'It is all mine but only as long as I am here, after that not anymore.'" For many Smith Islanders, these are stories connecting social ethics with sensory memories inscribed into places. When "wisdom sits in place" (Basso 1996), these spots have deeper cultural meaning for people who belong to these locales. As is true for so many Smith Islanders, playful times and sweet tastes from childhood conjoin with learning moral values, which evolve into local knowledge during one's life course.

Part of the land at North End has eroded, with hardly any signs of settlement there now: only a gravestone of a soldier from the Revolutionary War and tall old pecan trees. The sweet taste of the fruit from the old fruit trees is lost, along with Eddie Boy's memories after his death. Some people on the island still remember the stories about an old farm called Pittcroft, and when they recall memories of a former building there and the old orchard, they have a nostalgic glimmer in their eyes. Pittcroft, I was told, burned down long ago, but this farm, now invisible, still lives on through the stories embedded in the memories of older islanders. A place like Pittcroft is more than a nostalgic symbol from the past. To many, it is a reminder of the island's former fertility—of the fruit trees and foods with distinct tastes produced by the island's dairy farmers. One of the oldest photographs from Smith Island, taken in 1905, shows a man named Aaron standing next to a famous bull from the Pittcroft farm. Some members from an older generation remember Pittcroft as nourishing the whole island with its eggs, milk, and other dairy products. According to Junior, the Pittcroft farm had as many as fifteen cows. Pittcroft also served as a hotel and restaurant for visiting tourists. "It was a premium location on the Bay," Jennifer, who worked there, said with pride. Mostly, though, they remember the orchard and its many varieties of fruit trees.

The stories about Pittcroft sparked my interest in farming and gardening on Smith Island. It was not until I examined family albums that I could see more images of the islanders gardening. In addition, I came across multiple photographs of people posing for the camera alongside their animals: a child standing by a group of goats, or a more personal portrait of someone holding a hen. When I asked about the early settlers' work and sustainability, Eddie Boy told me that originally people came to Smith Island to farm and raise cattle, since seafood "didn't really take hold until the late 1800s and really early 1900s, because there wasn't anything like refrigeration or, well, people wasn't accustomed to eating it." To Eddie Boy, it seemed evident that people came to the island for its free property and to "have enough room to grow well." He did not remember when, exactly, the transition from farming to the seafood industry took place, but he had a childhood recollection of having known a man who supposedly shipped the first soft-shell crabs to the mainland and started the soft-shell crab industry on the island: "I guess staples of life of the day, because I can't see here on the island where it would be enough room to run a farm as we know it today."

Historically, island settlers practiced small-scale farming, grew a limited amount of crops for their own use, and kept a small number of domesticated animals. In addition, they developed traditions of hunting wild animals and fishing, many of which are illustrated in old photographs. In the nineteenth century, the islanders transitioned to a water economy, based on crabbing and oystering. What once was Pittcroft farmland is now covered with tall grasses sprouting from sandy ground. Parts of the land on North End have eroded and merged with the marshland, thus reducing the amount of available land for gardening. As the loss of land interfaces with losses of cultural meanings, aging islanders are destined to live with nostalgic memories of trees from their childhood experiences, and with the loss of a once-familiar soundscape, since the birds don't stay in a place without trees. The ghost trees are a powerful symbol of a death brought upon the island by its changing ecology.

When I returned to Smith Island in May of 2021, it was evident that the islanders had been working on land improvement by depositing sand to fill in multiple areas, as well as enlarging water channels across the island. From their current engagements with the land, the islanders are building up its resilience, using traditional land management practices. Nonetheless, I noticed that many of the familiar ghost trees I recalled from past visits were gone. They had fallen into the marshlands, and their skeletons were hidden in the tall grasses. Some of the remembered live trees had, over time, become new ghost trees. Focusing on the connection between the people of Smith Island and their land brings us to the intersection of several different modalities: shifting landlines, absent generational storylines, tree trunks with salty white lines, and disappearing kinship lines. When I asked about the land and water, Jennifer, a newcomer who had recently moved to the island from the Eastern Shore, gave a thoughtful response.

> Surface of the water is the same, but I always think about the bottom of the water. We think about the land under the water.

JERRY SMITH

When people say land, for us the land doesn't stop, for me land continues into the water. We are also thinking about animals that are in the water. You can float on the water on anything. Oh my, there are so many stories about land and water. Reflections at night on the water, ice skating in the winter, and we would do it at night, that's crazy, we know what is at the bottom. For example, when you see little things like horseshoe crabs, you know water is shallow.

As I listened to Jennifer, I recalled accompanying Eddie Boy on his boat as he deposited his crab pots in the water. I asked him about a little monitor attached to the panel in the cabin. He explained that what appears on the screen lets him know about the underwater seascape. Allan, a waterman who agreed to let me photograph him at work on the water, described how, while *dragging* (the term for harvesting soft-shell crabs), he imagines an underwater landscape. For Smith Islanders, landlines are always moving with the water's flow. While they manage the land in their yards and their waterfronts, they also know the limits to human control of the seascape. Accepting how the Bay's water constantly moves and shifts the island's shoreline deeply defines the Smith Islanders' existence.

4

Shifting Grounds

IRIS AND KEN EVANS

Iris and Ken

"I am ready, but it looks like the Lord wants me here," Iris greeted me. "I didn't expect to be here so long." She looked at me with her wide-open eyes and said, "I would give everything to be back on the island." Iris knows that when I visit her and her husband Ken in the nursing home, I am on my way to Smith Island. My visits seem to remind her of the loss of her island life. Iris and Ken, both in their nineties, left their home on Smith Island in the winter of 2016, when they became ill with pneumonia. They did not expect then that illness would take them permanently away from the island. Once they were discharged from the hospital, Ken was sent directly to a nursing home, and Iris temporarily stayed with her daughter Alice, who was also on the mainland, in Crisfield. When I phoned Iris, she was disoriented and frightened. Her voice sounded fragile and overwhelmed as she was trying to make sense of what was happening to her. She was confused about the future and worried about losing her shared life with her husband. She expressed how her longevity has presented her with many unknown challenges in later life.

I first met Iris and Ken in 1995, when they welcomed me and my husband onto their boat at the dock in Crisfield. As soon as the boat left the harbor, a heavy fog settled into the Smith Island channel. The invisibility of the waterway prolonged our trip for one to two hours. It was then that I learned Ken was a traditional waterman, and Iris managed a local store. When Ken discovered that I had grown up in Czechoslovakia during socialism, he asked me many questions about my childhood. Ken had been stationed in West Germany while serving in the US Army, a circumstance that contributed to our conversation lasting until he had docked his boat in the Smith Island marina. Upon our arrival, my husband and I discovered that the local B&B was closed, and there was no boat going back to the mainland that Saturday evening. Ken and Iris realized our situation and invited us, perfect strangers, to stay in their house.

When entering their home, I was struck by a sense of familiarity. Iris's display of Czech and German glass, which was on a shelf on the wall separating the kitchen from their dining area, was similar to the private interiors I remember from my hometown of

Susice in the southwestern Czech Republic. Next to the kitchen and dining room was a living room, with a sofa and armchair facing the television.

Iris served us dinner at 5 PM, their usual dinnertime. After dinner we watched television together, and later that evening, my husband and I took a walk through the town of Ewell. Smith Island houses were hidden in a dark, foggy night, and fenceless yards spread out all the way to the now-invisible marshland. The streetlights highlighted specific areas, but more distant landscapes were invisible. Several cars passed by, but we did not meet anybody walking on the road. When our eyes had adjusted to the darkness beyond the streetlights, we began to see cats emerging from the dark areas. They sat on the roads, rested in the trees, and lingered on the porches of the houses. In the presence of cats and the absence of people, it was as if we had entered a surreal world, resembling a *film noir*. The next day, we woke up to the smell of bacon and freshly made coffee. At breakfast, Ken and Iris spoke at length about their Sunday church service and their pastor, whom they loved. When I returned to the island in the summer of 2013, after my first brief visit in 1995, Iris was 89 and Ken 88. They remembered and welcomed me and my family into their home.

I stayed for a week that summer, and they introduced me to many other people from their community. I visited them almost every day that year and developed a stronger bond with them. In the following years, they invited me to participate in some of the traditional Smith Island rituals, many of which were connected to seasonal life on the island. I always sat next to Iris and Ken when I attended these events. Later, when Ken was not able to participate because of the pain in his legs, Iris continued going to them alone and still sat next to me. As her hearing became progressively worse, it became increasingly hard to have a conversation during these socials. She would sometimes reach over to me and silently hold my hand. Iris did not let her hearing disability get in the way of her connectivity to me or her community, however. She was always formally dressed, wore jewelry and makeup, and kept her hair cut short. She always looked stylish, yet unpretentious. "When I was 90, they [Smith Island women] made a birthday party for me, it was the first birthday party I ever had," Iris told me, with a bit of surprise lingering in her smile. Her ninetieth birthday party was her last social event on Smith Island. She left the island shortly thereafter and never went back.

After a brief period spent in her daughter's house on the mainland, Iris eventually moved to a nursing home. When I went to visit her, she was sharing a room with a stranger and living in anticipation of the next available opportunity to reunite with Ken. Eventually, Iris and Ken were offered a room in the nursing home where Ken was living, and this became their last shared home. The room was small, with only enough space for their beds and two armchairs, but the view from their window—of the waterfront, with the seascape in the distance—gave an expanded visual sense to their living space. Between the beds, which took up the left and right sides of the room, was a small open space for a visitor's chair. From their bedsides, Iris and Ken looked at a large, colorful quilt displayed on the opposite wall. It had been made from fabric printed with photo images of island life and the seascape. It was a gift from their neighbor Doris, an artist, who had briefly lived on the island. Next to the quilt was a display of family photographs, mostly portraits of their grandchildren. A television screen was hanging from the ceiling above the photos. The decorated wall in Ken and Iris's room reminded me of interiors in Smith Island homes. I noticed that every household has a designated wall or special display area for family photographs. The islanders typically frame some old family photographs and display them next to more current photographs of their children and grandchildren. In addition to freestanding framed photographs, many arrange the photos into a pictorial family tree on the wall. In some homes, small color photos fully cover refrigerator doors.

During my morning visits, I often saw the couple eating breakfast together. Iris was typically seated on the edge of her bed, and Ken would be in his reclining armchair. When I would stop by later in the day, though, Iris would usually be eating alone, and Ken would be in the dining room with his old friends from the island. Because Iris had lost her hearing, and her hearing aids did not work at all, she spent more time in their room than Ken. When she did socialize, Iris relied on Ken to tell her what was being discussed, because she could read his lips with ease. Over the years they had developed other strategies for mutual support in their daily tasks. When they lived on Smith Island, their joint compassion sustained their life there. After moving to the nursing home, Iris and Ken continued to care for each other while coping with their mounting disabilities. Smith Islanders joke about their codependent reliance: "When they travel on the mainland, Iris drives, but she can't hear; while Ken can't see well, his hearing is fine, so they drive together."

As soon as Ken and Iris had settled into their new life in the nursing home, each, in a unique way, reestablished their attachments to Smith Island life. Ken phoned his friends every week and discussed island life. Iris continued her everyday interest in reading the Bible, the only book she kept on her bedside table. Iris, like many other islanders of her generation, had a lifelong connection with biblical texts. In addition to reading her Bible privately, she and Ken had attended the Smith Island bible-study group in the church every Wednesday evening. Later, when she lived on the mainland, apart from her island community, she continued to read from her Bible and participate in the nursing home's church service. For many islanders, reading biblical texts provides a form of knowledge that can be shared with others during church services, storied in ordinary conversations on the road, and collectively discussed every Wednesday evening with a bible-study group. The Bible has a deeply personal meaning in people's daily lives on the island, and many read it every day. The cover of older islanders' Bibles is often worn from frequent handling, and the front page is usually inscribed with their family lineages, handwritten onto a predesigned family tree. Each Bible is personalized by underlinings in the text and short notes in the margins, and it is bookmarked with photographs or gift cards. For Iris, when living in a nursing home, the Bible was a tangible spiritual connection to the island community she left when moving to the mainland.

Ken compensated for his lost participation in island life in a different way. Having grown old during a time of major changes in communications technology, Ken and his fellow waterman used marine radios for enhancing their everyday work, as well as for the pleasure of belonging to a social network in the Bay community. Marine radios provide a constant means of communication between working watermen, their families, and their community across the Bay. Radio conversations connect people in boats, when they are on the water, with Smith Islanders at home or at work in their shanties on the island. Typically, but not exclusively, radio talk is related to work, weather, and ecology. There are also conversations about sports, politics, and news that are often spiced up by playful joking and teasing. Smith Islanders listen to marine radio talk while engaged in their daily lives. When watermen transition from full-time to part-time work, or stop working entirely, marine radios provide an important, unlimited connection with other watermen. When Ken lived on the island, he turned on his marine radio daily and always listened to the voices coming from the sea, even when he was involved in a conversation with Iris and me. For Ken and others from an older generation, marine radios represent engaging, available, social connectivity in their later life. It is something Ken still sought through phone conversations with islanders while he was in the nursing home.

Increased longevity presented Ken and Iris, as well as many of their fellow islanders, with more time initially to spend on Smith Island, and, later, finish out their lives in a nursing home, away from the island. A closer look at my conversations with Ken and Iris about their life in the nursing home shows their sense of lingering when entering later life. While they waited for death, they longed for their lost island ways. Even if Smith Islanders' relationship with death is somewhat normalized during their life course, and they became used to enduring the uncertainty island life brings to them, longevity, as we learn from Iris, creates a new type of uncertainty, stemming from a prolonged adulthood, and results in confusing feelings.

"There is a sense of life being fleeting," reflected Chris, a former waterman, when I asked him about Smith Islanders' aging process and their relationship with death. "We all know somebody lost on the water, a person who died young," he added. "My best friend, Dennis Evans, died in a boat explosion in March 1975. I was supposed to be with him, but couldn't go because of school," he said when I asked him about his memories of people who had died. "Leaky gas line killed him. The gas collected in the bilge under the cabin and when he lit a match to fire up the kerosene heater, the gas vapors exploded, killing him instantly." Dennis's brother Junior talked about that event during a boat ride on our way to a funeral, which took place in Tylerton: "It is getting worst with age for me. Older I am, more I think about that accident, day and night as I dream about it." There, on the water, it was clear to me from his voice, silenced sometimes by the rumbling boat engine, how greatly he suffered all his life from losing his brother Dennis to a horrible death. The longevity of his pain was written in the lines on his face.

Traditionally, Smith Islanders' relationship with death has been managed through storytelling and their Christian faith. Stories about death often relate to their knowledge of places and weather, and a belief in the power of God. I discussed this local knowledge in my early work (Kopelent Rehak 2019a, 2019b), when I learned about deaths on the water through my study of family albums. While I was collecting Smith Island historical photographs, I came across images and stories of people who died in a tragic accident on the water or from a sudden health crisis on the island. It is not unusual for the death of a fellow islander to become a part of childhood memories. "I can remember as children not being afraid of death. An example would be when someone had died and arrived at church the day before burial. We as children coming home from school would stop at church to see them in their coffin. No adults would encourage it or be with us," Mark responded when I asked him about his childhood memories of death.

Iris and Ken's late-life journey shows us how their membership in a small island community and its weather world continued to define their way of being as they inched closer to death. In the nursing home, distanced from Smith Island, Iris and Ken held on to any opportunity for a connection with their lost island life. The fabric of their life course was woven from threads of engagement with their community, their sense of self, and their sense of place. A feeling of loss, realized in connection to one's relationship with death, is embodied in one's later life on the island, but that loss is realized even more deeply when Smith Islanders must reluctantly move to the mainland. Ever-present losses are experienced through the reality of the weather world: the loss of land to water and erosion, the loss of their kin to death, and, possibly, the loss of the graves of the already-buried dead in the distant future. All of this is manifested through these multiple modalities of death in the fabric of the islanders' later lives. Living with a great sense of loss is an overwhelming, ever-present feeling for aging islanders. Therefore, in the next section I turn to the collective expressions of grief for lost kin manifested during mortuary rituals on Smith Island. I discuss the evolving history of these practices to contextualize the larger senses of grief in this community.

Expressions of Grief

Shortly after I moved to Baltimore in 1995, I saw a photograph of a gravestone in the front yard of an island house in the Bay. The photograph was part of a larger collection published in a book about Chesapeake Bay culture. I wondered about the implications of this front yard burial practice. I was curious to see the place where a gravestone would be the first thing a person would see when going out the door of their house. I made my first visit to Tangier Island and, later, to Smith Island, based on this interest. While I didn't find any gravestones in front yards on Smith Island, another opportunity for thinking about the islanders' relationship with death resurfaced in 2014, when Pastor Rick invited me to my first Smith Island funeral.

It was a hot, humid July day when a small group of women, along with Pastor Rick and church organist Clarence, gathered on the waterfront in the village of Ewell. The women, wearing their best dresses, ignored the intensive heat and humidity of the summer afternoon and chatted as they waited for the arrival of a funeral boat to take them across the water to the funeral in Tylerton. When the boat reached the harbor, everybody stepped on board. The trip to Tylerton was short, and the women continued to chitchat until we reached our destination. After landing, all of the passengers walked directly from the boat to the church. I didn't know many people at that point, since this was only my second summer on the island, but I recognized two boat captains, brothers named Larry and Terry. The deceased woman, Mina, was their mother.

My participation during extraordinary events on the island, conjoined with ordinary ways, like taking a boat ride together, has been very influential not only for the kind of data I collect during my fieldwork, but also on my writing style. Because of such experiences, in my narratives I can closely relate to Smith Islanders' ways of being.

At the beginning of the ceremony, the funeral attendees formed a line and walked up to Mina's sons and their families, who sat in the front row and faced the already closed casket. After an opening song and prayers, led by the pastor, designated people from the community spoke about their memories of Mina. Patty, in her fifties, was one of the talented storytellers and comic writers on the island, and she presented a series of humorous stories about Mina's life. Satiric comical performances during social events are very popular on Smith Island. In each generation and village, there are those who invent skits and write speeches to be presented at public events. Patty takes preparing for her performances quite seriously. She collects stories and writes her ideas on pieces of paper or in a notebook she carries when her family or friends socialize. She always has her notebook with her, even when she is at work, where she prepares food in a local store. When I asked her to meet with me and read some of her comic stories, she brought a bag full of papers, a collection she calls her "joke bag." Her language play, and that of others, is based on Smith Island social situations and experiences. At the funeral, when Patty introduced humor into her stories, the audience laughed without hesitation. Patty's narratives were received with a sense of ease, and the shared humor brought release to the grieving audience. When Patty finished, Pastor Rick invited others to share stories about their memories of Mina. After final prayers, the funeral ceremony continued in the graveyard. With a camera on my shoulder, I followed others to the gravesite. There I saw the three brothers, Mina's sons, standing side by side above their mother's grave and holding hands behind their backs. As I raised my camera and selected the shutter speed, some of the pallbearers turned and signaled that it was not appropriate to take pictures. Since then, however, I have observed that more people on the island are using their phones to take snapshots and videos during social events, including funerals. Cellphone photography is now part of all social events on the island, and it is assumed that performances at these events will appear on social media.

Mina's funeral ~~events~~ resumed at the social hall in the church basement, a typical setting for a traditional Smith Island dinner. Like other communal meals on the island, funeral dinners are well organized by the women. They divide the work when they prepare shared meals, which typically include pork, chicken, and seafood dishes, along with side dishes of dumplings, mashed potatoes, grits, homemade bread, and green beans. Communal meals start with a prayer, after which people serve themselves from the buffet table. Dinner traditionally ends with cake and iced tea. The women then clear the tables, wash dishes, and clean the kitchen. They work fast, have particular routines they follow, and coordinate their work.

Funeral rites are typically transformative expressions of shared grief. On Smith Island, grief is expressed in collective prayers and songs, as well as through personal storytelling during a traditional island funeral. Moreover, for Smith Islanders, funerals have a double meaning. They reconcile the loss of fellow islanders and, because many relatives return from the mainland for the funeral, they are often a unique homecoming event. Smith Island's funeral rites are painful reminders for both residents and their returning kin about the decline of the island's population and of its traditional practices. It is not nostalgia for the good old days that lingers during the funerals—since many know the old days are remembered as being very hard times, in terms of sustainability—but rather a sense of the deep loss of longevity in placemaking.

Chris, who left for the mainland, wrote about his views of Smith Island funerals and his feelings when he returned for them: "It is always sad. You are saying goodbye to a person, but you are also losing part of the past that is so important to you. Or life, how it used to be. There are so few of us left. Sometimes it feels like the last time we will all be together. Islanders believe strongly in an afterlife and had firm ideas as to what that next life will be. Basically, straight out of the Bible: streets of gold." Chris then went on to explain, "I will be going to my Uncle Ralph Ed's funeral next week, and it is going to be sad, unbearably so. I have no memories of Smith Island without him being there."

Sharing personal stories with others in a church during a funeral is an example of how islanders collectively create a social space in which people express and share their emotions for lost relatives and community members through their storytelling. Funeral events are balanced by an equilibrium between structure and freewheeling. The structure of funerals resembles that of other social events on Smith Island. In such instances, ritual patterns are defined, not only by raising a collective voice in prayers and in singing, but also in storytelling by the participants and their shared traditional dinner. During a funeral ceremony, Smith Islanders move with ease between formal and casual frames and mingle between these two parts of ceremony.

Frances Dize, a former resident, has written a book about Smith Island. In it, she describes how, in the past, when a person died, an island family traditionally would send someone to the church to "toll the dead—one stroke of the clapper for each year of life" (Dize 1990:127). According to Dize, the watermen working on the water would return home when they heard the bell. A woman would gently care for the body of the dead person until Aron, the undertaker, came from the mainland to collect the corpse. Dize emphasizes how some people remembered Aron's craftsmanship, making each coffin with care in what was called his "coffin house" (his workshop), a room attached to his grocery store in Crisfield. The deceased's body made a two-way journey in *King Tut*, the funeral boat. First, it was taken across the water to the mainland, for embalming in the Bradshaw Family Funeral Home in Crisfield. Then it was returned to Smith Island for the funeral ceremony.

The islanders' recognition of beauty in crafting coffins and respectfully embalming bodies is borne out in the conversations I have encountered when attending Smith Island funerals. At every funeral, they acknowledge the special care given to their deceased members by the funeral home, which is owned by a Bradshaw family who were originally from the island. For Smith Islanders, care for the dead is a gift of beautification. But now, due to the influence of modern medical and geriatric care, the gifts of beautification and "gentle care" (Dize 1990) performed by Smith Island women are no longer a part of the grieving process. Smith Island people will most likely die in a hospital or nursing home on the mainland. The beautification of an islander's corpse now solely depends on the professional work of the undertaker in a funeral home. The dead body travels from the nursing home or hospital to the funeral home, and then the corpse is placed in a casket and sent back to the Island for the funeral. The beautified body, in an open casket, is viewed for the last time by other islanders the evening before the funeral. The casket is closed for the funeral and buried in the graveyard at the end of the ceremony.

As changes in the trajectories of a person's dying shaped new social patterns of mortuary rites in the Smith Island community, the gift of care by one's relatives, both before and after death, has been lost. Duke, who originally was from Tylerton and now lives on the mainland, spoke of the loss of this gift of care in death as he was reading an old photograph of his grandparent's neighbors, Edwin and Nina.

> When I went up home, I always enjoyed running over to see what Edwin had caught that day netting. He was the kindest

man you ever wanted to meet. And Nina Ruth was sweeter than punch! Just really good people. One memory comes to mind. As my grandmother was dying of cancer, Nina Ruth would come every day and rub her feet to make her feel better. They were best friends, all the while talking about old times and taking her mind off the pains caused by cancer. My grandmother was the first patient on Smith Island to receive at-home hospice care, and they helped her stay on the island until her death.

Residents' living patterns have shifted from a multigenerational family structure to a single generation of older adults, who depend for care on a spouse or relatives in their extended family, and peer caregivers, whose own vitality is diminishing and who are often unable to fully care for others. Today, most islanders are moved into hospice care before death, and final caregiving for their bodies is not provided by family or friends. The result is loss of the Smith Island tradition of gift-giving for the dead.

Mortuary Poetry

The old mortuary tradition of gift-giving by caring for a dead person's body may not be possible today. Yet expressions of grief, as I noticed during funerals, translate into ritualized forms of collective poetics: in music and singing, in storytelling, and in poetry written for a deceased relative and read during the funeral rites. I conceptualize Smith Island's art of storytelling as a shared form of knowledge about their environment and their work. To do so, I focus on the poetics of the island's verbal art and further explore gift-giving through the poetry written by Jennings when his parents died.

The first winter snowstorm was beginning to cover Smith Island with light snow and ice in early December. It was 8 PM when I called Jennings and Edwina, and Edwina answered the phone. "You better come over soon. Television is not working after the

JENNINGS EVANS

snowstorm and Jen is all beside himself. It looks like he will climb the walls," she joked. Jennings greeted me as I stepped into their house. "I never imagined I would live so long," he said. He was emotionally distressed because the storm had disrupted their television connection. Jennings was a night owl, and I could see in Edwina's eyes that she was worried about keeping him calm. As I sat down, he said, "What do you do with yourself, when you can't do what you are used to?" Trying to be helpful, and secretly hoping for more of his storytelling, I suggested to Jennings that we could look at some of his family albums together.

Jennings was open to my idea, and Edwina insisted on making ice cream cones for us. While waiting for this treat, Jennings and I opened one of his albums and began to look at old photographs, which were placed next to news clips and his own writings. "This was my father and my mother," Jennings said, pointing at the portraits of his parents. "I wrote a poem about them, for my father and my mother. I read it at their funerals," he remarked when looking

EDWINA EVANS

at photos of his parents' graves. "I wrote it how I felt," he said about the two poems he created for his parents' funerals.

Jennings composed poetry over the course of his life and published it, along with his short stories and journalistic columns, in the local newspaper, the *Smith Island Times*. In contrast to his other writing, his published poetry was always presented anonymously, but these two mortuary poems were signed. When Edwina asked him why he didn't sign his published poetry, he merely smiled. It was as if he was ashamed to reveal his deep feelings to the public. Lila Abu-Lughod (2016:187) writes that poetry is "a discourse of vulnerability, expressing sentiments of devastating sadness, self-pity, and a sense of betrayal, or, in cases of love, a discourse of attachment and deep feeling." Poetic responses thus are distinct, formalized expressions of emotions that are forbidden in everyday speech acts, and ritualized poetry could be a form of anti-structure.

It was still snowing, and their television remained out of service when I asked Jennings if he would be open to reading his poetry to us. He initially said, "Oh, I am not very good at this, that was a long time ago, I don't know." With Edwina's encouragement, however, he eventually agreed to read the two poem he wrote for his father's and mother's funerals. Tears blinded his eyes as he was reading. His voice was raspy but tender as I recorded his poetic memories of the relationships he had with his parents.

MY FAREWELL TO MOTHER PAULINE EVANS
She never demanded me to excel,
But her kindly heart, always wished me well.
Her gentle ways, I can never forget,
Her friendly smile is with me yet.

I remember her patience, her gentle hand,
And how she taught me to respect my fellow man.
I will always be grateful to the Lord above,
For being a recipient of her love.

And now she parts from earth's strife,
To meet the master of eternal life.
To the greatest love that was ever known,
There's nothing more to fear, Mom.
You're going home. (Jennings L. Evans, 1988)

MY FAREWELL TO DAD CAPTAIN BEN EVANS
Time has closed the curtain
On his life here.

But I will always remember him,
His memory I revere.

He worked on the waters
Of the Bay and Sound.
He was ambitious, but honest
And not one to lead you down.

He had a sense of humor.
He was kind, and witty too.
But sometimes he would worry
When my boat was overdue.

Sometimes from the shore
His eyes would search the Bay
Until when he was satisfied
That I was on my way.

The Lord was good to him.
He gave him a long abundant life.
I've seen him on his knees in prayer,
Giving thanks in the morning and at night.

I'm going to miss him, in the lane
With his little doggie friend.
Farewell Old Captain, Captain Ben.

(Jennings L. Evans 1992)

Here, in his ritualized poetics of vulnerability, Jennings unveiled anxieties that permeate a desire not only to succeed, but also to survive unexpected, rapid weather changes. Both poems are unified by a sense of precariousness and then release in their imagined journey to an eternal home, a metaphor for belief in an afterlife that many Smith Islanders share. The gentle ways of his mother, expressed in her voice and by her hands—a patterned metaphor for vulnerability, embodied in anxieties over weakness—became part of Jennings's sensory memory process. Similarly, in the poem for his dead father, Jennings expressed an underlying duality in the core of their relationship. Defining his father's success, Jennings described him as ambitious yet witty, and humorous as well as humble, which brings us back to Jennings's ecological narratives, presented in chapter 1, where we clearly see his view of Smith Islanders as belonging to their weather world.

Jennings turned the album pages and remembered all of the people in the photos. Then he reached the pages holding photographs of his family's gravestones, etched with the names of his parents, his son, his grandparents, even his dog Kenny. I could tell from the color photos that this was not the current graveyard on the island, but an older one, hidden in the grassland. "I use[d] to go there, in my skiff. That was when I still could care for the gravestones, out past the North End," he explained. The gravesite where his parents are buried, shown in Jennings's pictures, is in one of the original settlements on Smith Island. Once connected to the main island, the old graveyard is now on a separate island, surrounded by marshlands. In the Bay, when wetlands take over the eroding land, but before water swallows all that remains, gravestones are the last survivors (Cronin 2005). Jennings's family's gravestones in the photograph, clinging to the eroding land, show us how, over time, people have sustained their island ways on such shifting grounds by migrating across the island. Like the hidden foundations and ditches found in old properties, which I discussed in chapter 3, the graveyards on the island are sites for reconstructing a past. Moreover, they reveal the flexibility demanded from people living on Smith Island, who have to relocate their houses from one area to another on the island. Given that flooding seasons are getting longer, resulting in increased land erosion, the burial grounds in coastal lowland communities like Smith Island have become unstable, shifting sites.

A Final Gift

The evening before Ken's funeral, Hoss, the gravedigger, was finishing his work. As I walked closer to the graveyard, he casually said, "This is the last thing I can do, and somebody will do it for me one day." Watching Hoss shoveling up dirt, I thought about how his voice gave a sense of the purpose of gift-giving. I first met him the previous summer, in 2013, when I had come to Smith Island with my family and rented a house for a week. Hoss walked by the house in the late afternoon and engaged my husband and children in conversation while they were in the front yard. I was new to the island, a stranger, and Hoss impressed me as being a giver. The next day, he came by with a bag of figs from his garden. The following summer, in 2014, Hoss continued his gift-giving and often left a bag of figs on my doorknob before sunrise.

Later on during that week in 2013, when he invited us for a visit to his house, my husband and I came to know more about Hoss. His home is just on the edge of the marshland. It is a three-story wooden house on the side road in Ewell, one of the oldest houses on the island. It was a gift from his long-dead friend Reuben. Reuben had been a resident artist on the island and was a liked outsider in the Smith Island community. His realistic paintings depict working watermen in the seascape and are part of the artwork displayed on the walls in island homes. In addition to purchasing Reuben's paintings, some watermen took art classes from him. When Reuben died, Hoss inherited the house, with its painting studio, and he began to paint. Hoss told me that by watching Reuben, he learned about colors and composition. Hoss's paintings of seascapes and life on the water are similar in style to folk or naïve art and differ from Reuben's realism.

Stepping inside, I could see that Hoss's home was hardly habitable—a dilapidated fragment of a human dwelling, reminiscent of a partially constructed theatrical stage set. It was obvious that Hoss had no means to care for the house, yet the gift from a deceased friend seemed relevant through its association with a creative heritage. Hoss's survival on the island depends on odd jobs and service to the community. In addition to digging graves, he helps with daily food deliveries from the boat to people's homes and pitches in when there is a need for extra hands. Through his service to others, Hoss sees himself as an artist, and also a giver, continuing this practice through gravedigging.

Hoss's life took a new turn when unexpected illness altered his health. During my stay in the spring of 2021, I saw Hoss, dressed in short pants and a long tee-shirt, as he was walking home from work. When I asked him how he was, Hoss lifted his tee-shirt, revealing a vertical cut across his stomach. "I have prostate cancer, and this is what happened, but just today I talked to my doctor, and he wants to see me in six months. I must take it easy, and so I retired from grave digging and can't lift anything heavy but I will live," he declared as he was walking away from me. Our conversation echoed across a dirt road leading up to his house and into the marshland.

5

Broken Bodies

MARK KITCHING

At an island house party one Saturday night, I met a young man in his twenties who had recently decided to work on the water. His friends at the party were excited about his decision. Since fewer young watermen are pursuing traditional work, I was curious about his commitment to it. When I asked him, he smiled and explained,

"Well, I will try, and if by the time I am 40 my body is not broken, I am good." This vision of his future, contemplating a broken body, left me thinking about watermen's ways of growing old and their work.

"Watermen always have health issues due to their work. Loss of hearing is one because of the constant roar of the engine. Knees and back problems are another one. Eye problems from the sun. It is hard work, and it wears a body," said Mark, one of the watermen. Another of them, Jerry, added, "My wrists are bad from the years of [crab] potting, I have carpal tunnel and sometimes I can't sleep they hurt so bad." Listening, I was reminded of a passage from *The Old Man and the Sea*: "The old man was thin and gaunt with deep wrinkles in the back of his neck. The brown blotches of the benevolent skin cancer the sun brings from its reflection on the tropic sea were on his checks. The blotches ran well down the side of his face and his hands had the deep-creased scars from handling heavy fish on the cords. But none of these scars were fresh. They were as old as erosions in a fishless desert" (Hemingway 1952:9).

Working out in the open, on the water, weathers a man's face. Sun burns the skin on their arms, even if the watermen use sunscreen. Their hands are swollen and often cut from their labor. During the crabbing and oystering seasons, exhaustion from hard work every day is part of being waterman. While their minds may be used to harshness, their bodies are generally weakened from being continuously exposed and, often, injured. The Smith Islanders' health trajectories are a focal point of everyday life in this aging community. In this chapter I discuss how watermen's bodies are weathered by their work, and how their subjectivity, illness, and pain are intertwined with the social dimensions of health.

"The body adjusts for the most part. However, when working in rough weather you are often times off balance when lifting or overreaching, or whatever it takes to get the task at hand done," said waterman Mark, reminding us that his experience is a "double motion" when at work in the boat (see Dalidowicz 2015). Later on in our conversation, Mark jokingly admitted, "When you are young, you don't notice or talk about it, as you and others work hard. When you get old you just moan about it. When I go to the doctor, most of the times I forget to say anything about it." His internalized chronic pain has grown into everyday silence. "Everybody was so busy, and so nobody talked about it," Mark said when I asked him how much his father or grandfather talked openly about the chronic pain from their work. Lindsey Martine's (2013) discussion of how health treatments can effectively address disruptions of one's life course shows how people reinvent their narratives as they make sense of their pain at various points in their life course.

While women in this community do not work on the water, their traditional labor on the island is also physically demanding. Their jobs—tending to the floats containing soft shell crabs and picking the meat out of hard shell crabs—are repetitive and physically strenuous. During the crab-picking season, when women are sitting and plucking out crab meat for long hours, their shoulders, arms, fingers, and lower backs are in constant pain. Pain has long been recognized by doctors, social scientists, and philosophers as a human condition that is difficult to express. Elaine Scarry (1985) writes about the hidden voice of physical pain. Yet, when it is finally heard, it begins to tell the story. In my research I have listened to various narratives about life with chronic pain from work, addressing situations where this voice is suppressed. The chronic pain that Mark and others endure indicates a silent agent in a body pushed beyond its limits and subsequently broken. When I asked Mark how he thought pain is managed, he said, "I don't know if I have learned that yet." He then went on to explain what he meant.

The pace you set usually allows you to do what you have to do that day. The job I do allows me to work 'til a certain time or, [when] oystering, a limit has been achieved. I was fishing crab pots and had 500 pots in the water to tend to daily by myself. I could not do it. However, I could probably tend to 200–250 pots a day. At the end of either of these days, I would be very tired. It is part of a skill, because you can't exceed these limitations and successfully work on the water.

Long-term exposure to heat while working in humid marshlands also pushes personal limits. Once I was walking by John's shanty in Rhodes Point after he had returned from a day spent at sea. Our conversation led to narratives clearly defining how exhausting crabbing can be during the summer months: "One morning, I got up and as I was putting my shoes on, I fell asleep while bending over." Lines are etched on John's face from an exhausting battle between heat from the sun and his own degree of vitality.

WORKING ON THE WATER

"I did have a skin cancer removed from my face on two occasions. It was a couple of very lean years crabbing, so we opted to take that summer off from crabbing, and I worked with John on some renovations for the fire department," Mark told me when I had asked him about his exposure to the sun.

> Sunshine and heat are necessary for a good crab season, so you adjust. Work at a pace you can handle. You put an awning overhead to have shade. If you have a slight breeze of five to ten miles per hour on a hot day, it is usually more comfortable about 11 AM than at 8:30 AM. At that time the sun is overhead and complete shade is covering you. You then have to contend with heat when arriving at the crab house to do work there. At the end of the day, about 1 to 2 PM, you run out of steam. At the end of that, I have trouble concentrating and remembering work I must do when I get home. Even making a phone call I will occasionally put off.

Tracing Mark's steps enables us to see how he can manage his bodily strength and chronic pain by using a particular way of knowing and recognizing his limits. While Mark and John had to take one summer off from crabbing because of cancer, since then they have recovered and, like all waterman, returned to their work on the water the following season. Smith Islanders take a minimum amount of time to recover from serious illness, and they return to work shortly thereafter, but they keep silent about their minor injuries.

When I walked across the island taking photographs, I often saw watermen coming back from a workday on the water. Some wear long-sleeved shirts while fishing crabs, but all watermen cover their faces with a thick layer of sunscreen and wear long pants, which protect them from green and brown flies. Roadside conversations with these men often revealed how heat and humidity impacted them. In addition to long-term chronic pain from injuries and repetitive manual work, one of the most common hardships affecting the health of people working on the water is their exposure to weather, particularly the sun. In the winter, strong winds make the water rough, and in summer, heat from the sun and humidity in the marshlands take a toll on exposed bodies. Besides the oppressive heat, the sun affects watermen's health in two ways: sun exposure, resulting in skin cancer, and bright light on the water, contributing to cataracts.

August is a breaking point, the peak of heat in the summer season. That month in 2018, "Jessy Boy" was walking from the boat toward his home when he met and greeted me: "It was so hot out there on the water today, I just couldn't take it anymore and had to take my long-sleeved shirt off. I just worked in my white tee-shirt." Jessy Boy's face was covered with a thin layer of white sunscreen, and, like most of the watermen, he wore a baseball hat to shade his face. I could see exhaustion in his eyes from a day's hard work on the sea.

It reminded me of days when I went out on the water to photograph and study watermen's workdays. The sun was invasive, but a

JESSY BOY BRIMER

simple cloth draped over a small area of the boat provided shade, which seemed life-changing at the time. Seeing Jessy Boy and other watermen coming home from a hot summer's day of work at sea brings back memories of my experience and how I felt returning to the island after a hot, humid summer day on the workboat. I could barely walk back to the house where I was staying and, once there, I collapsed into a deep sleep for two hours. But my own experience of a day on the water gives me only a small glimpse into how, year after year, heat can impact a person during a long summer crabbing season, followed by a winter oystering season.

As watermen grow old, they must learn how to manage their cumulative chronic pain. Eddie Boy talked about "pacing [his] days, or weighing [his] time between work in the sea and shanty" in his later life. It seems that as they age, Smith Island watermen continue to master the limitations of time and body, a skill they have developed during their life course. Many realize increasingly greater limits to their physical strength, yet they do not think about retirement. Rather, they speak of changing speed, or "changing pace," as Eddie Boy put it. "Most I fear myself. I fear a day when I discover that I don't have it in me anymore," revealed Morris when an outsider asked him if he ever feared the sea and the weather. In their silence, watermen acknowledge their broken bodies and are fearful about their diminishing vitality, even as the heat of the sun is inscribing lines, born of years on the water, onto their skin. Ernest Hemingway (1952:10) aptly describes how the weather is written into the face of someone who works outside in the open: "Everything about him was old except his eyes and they were the same color as the sea and were cheerful and undefeated."

Caring for Broken Bodies

"People are living longer here too, yes. People always kept working, but some just didn't live so long. Knee replacement and heart bypass

MORRIS MARSH

surgery has helped a lot. Also, not as many people smoke anymore," Mark confirmed during our conversation about longevity on Smith Island. While working on the water is demanding, most older adults on the island who engaged in the traditional work of crabbing and oystering live and work much longer than previous generations. These older adults on Smith Island continue to have active lives as they age, yet they face new dilemmas. For current islanders, it is undeniable that they perhaps are, as Eddie Boy emphasized, "the last generation of Smith Island watermen." In a community that depends economically and socially on the participation of all its members, care for broken bodies, as well as preventive care and healing from sudden illness, is a communal obligation as well as a personal necessity. While working, older adults try to balance meeting their economic needs and maintaining their collective heritage while coping with a loss of individual and collective vitality.

Actively aging on an island where most people are involved in physical work demands a flexible attitude toward adjusting the speed of and time spent working when dealing with one's changing health in the process of growing old. To sustain their island life, residents must monitor their well-being and utilize available medical care from their health care providers. Historically, primary health care was supplied by a doctor or, later, by a nurse in residence. People still refer to a white, two-story house in the center of Ewell as "the doctor's house." Mary Ruth and Elmer Jr. recalled some of the resident doctors.

> Dr. Gentry was here during the 1960s. He married a beautiful lady from France and she came with him. She eventually moved away, but he stayed here. He was a good friend with Char and Ruke, and at the end of his life he stayed with them. He was a colonel in the army during World War II. He and General Westmorland, the general of the National Guard, were best friends, and he came to see Dr. Gentry. In the '50s Dr. Stout made house calls, and [later] Dr. Chambers and his wife, then Dr. Hefner, before Dr. Gentry. Dr. Shim Becker was from Australia, a certified nurse. Islanders loved her. She was married to Reuben, the painter who lived here. When some local guys studied every night together for a captain's license, Shim studied with them and got hers, too.

Chris, a former waterman, also responded when I asked about resident doctors.

I have a story about a doctor. His name was Thomas Gentry and he was an army doctor before he came to live on the island. He was a part of a group known as Flying Tigers from World War II. After Pearl Harbor, they were integrated to the US Army Air Force, Gentry as a doctor. He was the chief flight physician many years after the war. At some point he was married to a French woman, and they had a daughter. She used to come visit him on the island. She was a young and beautiful French woman, I use to gawk at her, she was stunning thing to see on our little island. Dr. Gentry died on the island and is buried there. Before he died, he needed help. Char and Ruke took him into their house and cared of him. When he died, the funeral on the island was big, a performance the island never saw before.

Mary Ruth's accounts about the resident doctors on the island were brief, storied memories of medical practices. They registered in her narratives as social knowledge about the integration of doctors who were outsiders to the island community. These tales not only reveal how newcomers engage with the Smith Island community, but also how the islanders view their own commitments, their previous experiences, and their role in integrating new settlers into life on the island. Their health care narratives show how medical practice on Smith Island was a limitless service, a commitment that overwhelmingly took time away from the doctor's personal life. "When something happened, we would just call to the doctor's house and he would come to our house; 'doctors' calls,' we call it," said Maxine. "He would go from house to house and check on people. There was no real office hours for the doctor, and when there was an emergency, you got the mail boat to take you." I met with Maxine in the fall of 2018, because I was interested to know more about her grandmother Lynette, the island's midwife. After I stepped into her living room and noticed family photographs on the walls, Maxine pointed out one frame that displayed several historical photographs. She stood in the middle of the living room and identified the people in the images as her grandfather and grandmother.

> I remember my grandfather's death. That is all I can remember about him and his wife Lynette, a midwife for the island. I think she liked to take care of people when sick, from what I remember. When doctor was here, he knew she liked to take care of people and so he would call on her to come and help. I don't know if she learned on her own, but I would love to know. I had her nursing bag here, stethoscope and needles, and when Hurricane Sandy came it was in the shanty and ruined. There was even a bottle with medicine. I remember when people went to Crisfield [for medical treatment] with them and stayed with them. People would go to her to get stitches, instead of going to Crisfield.

In the absence of a resident medical doctor, Smith Islanders must travel forty-five minutes by boat for health care on the mainland. The journeys to doctors pose a hardship for the islanders in many ways. Every trip takes up their full workday, and they find it to be costly. After witnessing these people's hardships when they become ill and are forced to frequently travel to the mainland for treatments, I was enthusiastic when I heard about the telemedicine program planned for Smith Island. I was on the island when a medical representative from the mainland, Carrie Pelican, came to a Smith Island United meeting in the summer of 2017 and introduced telemedicine as a means of providing long-distance care. Over the next few years, between 2017 and 2020, I heard brief comments from people about it during casual conversations. I sensed an overall hesitation in their voices, related mainly to the fact that they did not want to change their primary care doctor in later life, and their current doctor would not participate in the telemedicine program. Many islanders find their doctors through island friends and families and have established relationships with their health care providers. I could see how making a change is not an easy or a comfortable proposition for them. The telemedicine practice's office was first established in the small freestanding schoolhouse next to the church in Ewell. Janet, one of the middle-aged residents, received training in how to provide online administrative support for the islanders and schedule telemedicine appointments.

In the spring of 2020, I conducted a phone interview with Dr. Reynolds, one of the founders of the Smith Island telemedicine program. We spoke during the first wave of the COVID-19 pandemic in the United States, when distanced care in aging communities had gained a new meaning. When I asked him why he selected Smith Island, he explained, "It was a very important place to me, because it proved to be the single most geographically isolated place in Maryland." During our call, Dr. Reynolds explained in detail how difficult it was to establish a telemedicine program on the island. He described his original proposal for the program and spoke about the multiple challenges he encountered. "It was very difficult in every single way. Nothing about it was easy," he told me. Dr. Reynolds outlined three distinct entities that complicated the process of establishing a distanced care program for Smith Island: legislation, providers, and the island community. "First we had to overcome the problem with reimbursement for alternatives to face-to-face care," he stated. Dr. Reynolds then discussed the struggles he encountered with health department rules and regulations covering distanced care and patient welfare, noting the argument he used to support the original proposal: "When I teach medical students the history of medicine, I emphasize to them that diagnosis depends mostly on two methods: personal medical history narratives and a doctor's observations, both possible to practice by long-distance care such as telemedicine." Dr. Reynolds's description of his experience with established health care legislation fits in well with the politics of medical care in the United States. Generally, health care coverage boundaries are established to place protection for corporate profits ahead of provisions to give satisfying care for people in need. "But after all, the single biggest obstacle was to find providers," he continued. "With the help of Carrie Pelican, I was able to get grants for technology and established the center in Crisfield; she was the main administrator. It was a gift to the island, but when Carrie moved, as she was invited to work in one of the biggest telemedicine programs in the country, in Utah, it was difficult to find a place that would take the program. The community was not interested to take it to the island."

The process resulted in a collaboration among the medical group, the program's leadership in Salisbury (a larger city in the region), and the Smith Island community, which provided office space on the island and an administrative assistant, chosen from the island's residents. "The community eventually came around, but health insurance continues to be problematic. Doesn't matter what we do, just when we think that we solved the problem, they find ways to benefit [themselves] and complicate the situation," concluded Dr. Reynolds. At the end of our conversation, Dr. Reynolds returned to his diagnostic teaching philosophy vis-à-vis telemedicine: "Because the doctor is always looking at the patient and gaining the verbal history by talking with the patient, it is like being there." Following our online meeting, he emailed, "If you can show what it is like being there, that would be very helpful for the future of telemedical distant care."

Dr. Reynolds's account of the process of establishing telemedicine in a particular locale raises an important question about the complex politics of medical care in a place like Smith Island.

Following his email message, it seemed important for me to investigate how the practice of telemedicine is received by Smith Islanders. It was Maxine who first responded about her experience with this new medical program.

> I used it one time and I thought it was nice that I didn't have to go off that morning on the boat. She came to my house and set it up, and a doctor came on. She asked questions and then sent me my medicine. I thought it was great. It was a new doctor, not my doctor, but it was not like being there, but it was still good to have it. It keeps me from going on the mainland. I saved some money for doing the telemedicine and I did get help. I sure like it. In addition, we don't have to go to the mainland to get blood work done. It is done here and that was wonderful, we are thankful for that.

To fully understand the potential of telemedicine for Smith Island, we must realize that many islanders dislike travel to the mainland, and when they "go off" (leave the island), it is often because of a medical necessity, such as when they are ill and need regular medical treatment, in addition to their annual checkups. For aging Smith Islanders, traveling to the mainland is expensive and complicated, as they must arrange additional transportation from the boat to their final destination. The telemedicine program not only lowers the cost, but also eliminates the stress caused by traveling to the mainland. Kathy, who suffers from PTSD, spoke about the benefits of long-distance therapy: "I use telemedicine every week for my therapy. I don't mind using a computer; it has relieved me from the stress and anxiety of having to go off the island. When COVID started, the Veterans' Administration was right on point by using telemedicine, and plus I didn't have the expense and worry about catching the boat home. You don't have to push yourself to go, and you don't have to feel exhausted from taking the trip." She further explained, "I am kind of early [retired], I don't deal with stress well. I was in the military, and I do have PTSD and all that comes with that. In response to stress, I have had two times a stroke. My doctor told me that I have to slow down if I want to live."

Kathy and her husband reside in Rhodes Point. After searching for a peaceful place to live, Kathy moved to Smith Island hoping to manage her health conditions—"to stay alive," as she put it. "The island is my sanctuary," she said when I asked her if she would consider it to be a healing place for her. "Literally, this island saved my life. I love it. You can't beat serenity and community spirit. It is heaven on earth here," she claimed. She then continued to tell me what living on Smith Island meant to her: "Everybody knows everybody, and everybody helps everybody, like in a small town when I was growing up. In that sense we are behind, we are on our [own] clock. When something happens in your family, it is understood by the community that you take care of your family, and somebody will fill in. This is perfect; it is a wonderful place to be, for a family to raise children and to retire. You can feel safe and don't have to worry about any crime." She then exclaimed, "Look at it, from my window," pointing out the view from her studio room. "I get to see this view every day and it changes all the time." She paused and looked at me as she asked, "Where else can you see this water view and not see cars and condos?" She paused again, and her voice changed when she calmly added, "It is absolutely perfect." Then she continued reflecting on the value of this community for her.

> It is truly unique here. When we know somebody is not doing well, somebody will come by and help. For example, when we know Doris is sick, somebody will come and take [out] her trash. I moved here because of that, because of the community. Islanders are not reclusive; this is a real working community.

You have to understand crabbing is a back-breaking job. I think many are afraid to stop because if they stop, they may die. Many are in their seventies, and the oldest waterman is 80. I believe 70% the population here is over 60; the last young one is already crabbing and so pretty soon we will lose a way of life. Last year three older men had to sell their boats. Milty is in his sixties, he has some mental challenges, but he has a job here, he goes where he wants, and he knows when he has to be home. He could never have such freedom any other place, on the mainland. I couldn't walk on the mainland on the streets. He is creative, he collects things and makes boats.

Kathy's narratives clearly show how a person's sense of healing can be constructed through curative properties manifested in one's relationship to the natural environment, as well as in a social life expressed through collective compassion for others. In contrast to the watermen's bodies, broken from hard work on the water, life on Smith Island offers healing for a newcomer like Kathy. In this sense, her narratives about healing in place bring our attention back to Jennings's perspective on the island's habitat and the related concept of a double bind, this time in relation to lived health experiences. Kathy's mental health derives from her multiple levels of engagement with the island's habitat: her creative practices, her sensory observations of beauty and sound, and sociability within the community, which I discuss further in chapter 7. Environmental conditions impact the health of working Smith Islanders in major ways, such as in the life history of Eddie Boy, discussed in chapter 2. We see there how the islanders depend on nourishment from their natural environment and their community. For them, ecological and environmental sustenance is manifested in familiar aesthetics and sensory joy in everyday experiences. They are aware of this, yet they seem to protect their privacy—the secrecy of their place—

behind a veil of silence. In contrast to Kathy and other newcomers, native Smith Islanders normally do not talk about healing or the magical properties of their island. Instead, they derive nourishment and satisfaction from their environment, as enmeshed with the island's social life and embodied in the strength they identify as part of their heritage. I return to this concept in chapter 7 when I discuss notions of the island's future.

Pain of Others

"There is a lot of illness going around," said my friend Peggy when I called her. Peggy is a widow and lives alone. She has two grown sons who both moved to the mainland, where they live with their own families. She cared for her sick mother until her mother's death, and she took care of her husband before he died. In her house, on the wall in the hallway between the living room and kitchen, is a gallery of photographic family portraits. The color images of her husband, children, and grandchildren, arranged by her husband in the shape of a family tree, tells the past and future story of their family. This portrait gallery on the wall has an important place in her older life. Until recently, she had participated in the lives of her sons' families, as well as in the lives of her friends and community. Her current health condition, a kidney dysfunction requiring daily dialysis, prevents her from having any active social life.

I met Peggy, who is Iris's niece, in church. Whenever I came to the island, I talked to Peggy every day, either in person or by phone. Sometimes we would share dinner, and Peggy often reminded me that I was welcome to stay in her house. When I first met Peggy, she was employed at Ruke's Restaurant, but she stopped working there when the pain in her legs increased. Despite the suffering from what Peggy called her "bad knees," she then turned to helping others by picking crabmeat during the crab season. Ever since she was diagnosed with a kidney dysfunction, she has been closely monitored by her primary care doctor. "I so dreaded it when I have to go off to the mainland," she confessed, regarding her regular checkups. "I am in so much pain after the days I have to go there. Night before I go off, I can't sleep, and when I get to the mainland in the morning, I have to walk to my car first, before I can drive to the doctor. By the time I am done and come home after all day, I am exhausted." As I was getting to know Peggy, I noticed consistent patterns in our conversations, both about her health and that of others on the island. The subject of long-term illness, or news about a sudden illness in the community, had been a focus of our conversations, whether in person or on the phone. Like so many others, Peggy spoke openly to me about her health, as well as the illnesses of others.

The illness, pain, and suffering of others becomes collective knowledge, not only shared in conversations and greetings, but also expressed in prayers and faith-based healing practices in the church. Thus these difficulties enter multiple levels of connectivity as a register of pain. Beyond being part of casual conversations, either in person or on social media, anxiety regarding the pain of others is expressed during "socials" (the term Smith Islanders use for their social events) and church meetings. Through prayers, announcements, and collective healing practices, sharing suffering due to sickness extends beyond narratives of pain. The healing process involves people in the community in multiple ways. Smith Islanders' compassion and the depth of their care for fellow islanders—whose bodies might be broken from work at sea or are recovering from a sudden illness—surfaced often during church events. "I have seen watermen catch crabs for another waterman who was broken down," Pastor Everett reflected during our conversation about islanders' health and the scope of caring for others on Smith Island. "In church, people will come forward and ask for prayer for individuals going through a hard time," he said. He also explained more about healing practices in the church: "They will stand in on behalf of the person. For example, Carol Ann will stand in for her mother. We will lay hands on Carol Ann and pray for her mother."

Similarly, Sheila reflected on the time when she was undergoing intensive chemotherapy: "It was worth a lot to have my community there to help in any way they could. Whether it be to pray for me, send cards, visit, bring food, or just stop by to talk. I feel so blessed to live in a community that pitches in and does whatever is needed to make life easier." From Pastor Everett's and Sheila's examples, we see how Smith Islanders relate to others' illnesses by embracing the pain of others and participating in the process of healing. In our conversation, Pastor Everett also emphasized how, for this community, "sickness affects everyone." He compared a family-like structure within one household to the structure of the entire community.

> You probably have already heard this, but sickness affects everyone. If you think of a family in a home, mother, father, children, when one gets sick, everyone in the house is affected. The mother may need to stay home from work. The kids may need to help out with chores. When someone on the island gets sick, others step in and help. One family that was struggling, out of work for a year because of health problems, had the community bringing them dinners, paying their bill to the store, and giving them money for boat fare. One neighbor, after cutting his own grass, came over and cut their grass just to help them out. Everyone is constantly asking each other about updates of those that are sick. People call or post on Facebook for prayer chains.

Pastor Everett's narratives reinforced my own observations of Smith Islanders' responses to the pain of others. Social compassion for others in pain, whether through prayers or practical help, is striking when seen through the eyes of newcomers to the island. Precarious working conditions on the water and the great distance from health care on the mainland are major obstacles for aging islanders. Yet many seem to find their own well-being from the collective compassion elicited by a health crisis. It was my friendship with Peggy that allowed me to learn about changes that happened after World War II, when work on the water expanded from the predominantly winter-based oystering economy to crabbing during the warm season (from April to October), when they dredge for soft shell crabs and catch hard crabs with crab pots. This shift has dramatically impacted the day-to-day lives of Smith Island women. During the crabbing season, men work in the marshlands from early morning until midafternoon. Both men and women tend to the soft-shell crab floats, which they maintain in shanties, while the women will often spend time steaming and then picking the meat from hard-shell crabs. Women have learned their crab-picking skills by observing the previous generation of women who worked in pick houses. Picking allowed me to see the social dimensions of illness on the island. Throughout Peggy's and my conversations and her interactions with others, I could see how the pain of another person is embodied in the ordinary narratives recounted every day. It was then that I began to notice how Smith Islanders' personal pain and illness enter into public discourse. In addition to publicly expressed anxieties regarding health, narratives of illness and pain circulate in the Smith Island community, not only in casual conversations on the side of the road, during boat rides, or on the phone, but also on a new social platform, Facebook.

Peggy's life changed drastically when her kidney condition worsened. Because her daily dialysis treatments are time consuming, her advanced kidney disease now prevents her from working at all. Although able to remain in her home, Peggy worried at first about the prospect of no work and its impact on her financial situation. Peggy's treatments not only limit her participation in social life in person, but also shorten the time she has for connecting with others over the phone. Since she lives alone, phone conversations have always been important for Peggy. Before her health began to worsen, she often talked on the phone to her family and friends and had been active in community life on the island.

When I visited Peggy in December of 2019, she seemed to have adjusted well to her new daily self-care routine, despite limits to her full participation in a normal social life. As I walked into her home, I immediately noticed a change in her living room. The sofa, where I normally would have sat, was covered with a new blanket, a gift from her younger son's family. It was not an ordinary blanket, but one with an enlarged color photograph, a group portrait of her son's family. Peggy invited me to sit down on the sofa, but I could not bring myself to sit on the faces printed on the fabric of the blanket. The photoprint fabric so well simulated the presence of Peggy's kin that the pictorial illusion made me hesitant to violate the imitative magic of the blanket by sitting down there. It was a powerful reminder of how Smith Islanders integrate family photographic portraits into their sensory living in both their personal spaces and the interiors of their homes. In this community, losing the oldest islanders to illness and death is managed by a collective sharing of pain. Losing children and grandchildren to life on the mainland has resulted in different forms of adjustment to the solitude and silence of an empty home. While aging islanders travel to the mainland for holidays, during their ordinary days when living on the island, their connectivity with kin depends on phone calls or Zoom sessions. Therefore, in this daily solitude, the photographic portraits of their relatives—whether framed, hidden in albums, or printed on blankets—hold a promise for them in their later life. At Junior and Mary Ruth's house, I saw a family photo blanket designed as a multigenerational photo collage. This collection of family portraits

PEGGY'S HOME

included a sonogram image of their future grandson, who would be born on the mainland. It is through these photographic images that Smith Islanders hold on to their family's lineage and kinship bonds.

Family photographs acquired new meaning for Peggy during the COVID-19 pandemic in the spring of 2020. Because of her pre-existing condition, she was at high risk of getting an infection and therefore was forced to live through even more-severe isolation. Because of her engagements in community life, which were always central to her life tasks, I was interested in learning how Peggy was managing her isolation and inability to work. I visualized Peggy at home, in solitude, resting on the sofa between her daily treatments, covered with her new blanket. In my ethnographic imagination, I pictured her surrendering to the imitative magic embodied in the photoprint blanket, offering her a sense of connection with her son's family, even though they live at a distance on the mainland. Peggy, like most aging Smith Islanders, travels to the mainland and spends a limited amount of time with her children and their families during holidays. Most of the current islanders who are similar in age to Peggy's children are newcomers, often in their midlife. As many of them replace the last island-born generation, older people must rely on help from remaining relatives or these newly arriving neighbors.

6

The Taste of Things and Comic Relief

CHARLOTTE "CHAR" DIZE

Humorous Knowledge

Char was resting in the armchair in her living room when I walked into her house. After Ken and Iris moved to the nursing home, she and Jennings became the oldest people on the island. I lived not far from her house, on the same road, and I came to know her well after her husband Ruke died. When I walked by her house with my daughters, she always engaged us in conversations and often handed candies to my kids. I could see she was friendly and welcoming, but also lonely and still grieving for her husband. My daughters became regular visitors at her home. She loved their company and was generous with soft drinks and sweet treats. When my daughters helped her fix a disconnected television, it furthered our neighborly friendship. I often came to her house unannounced, and she would always welcome me in. During these spontaneous visits, we watched television together while sipping iced tea or soft drinks. Char remembered stories about her late husband and the time when she managed their restaurant. When I began my Family Frames project, I asked Char if she would be open to talking about the photos displayed in freestanding frames in her living room. Char reached for the photograph in a freestanding frame on a side table by her armchair, depicting her with her mother in the restaurant they ran together, and reminisced about that time.

> We all worked together, very much. We had some good times, too. It was always very busy, rushing, because it was only place. We started together, me and my mother. You see my dad just died, and she was widow, and the store was for sale. I said to her we had to become partners. I told my husband and he said, 'You can buy but don't depend on me, because I will not work there.' Longer we had it more he loved to come there. It grew on him. Men would come there and talk about things; the porch would be full in the summer. And my mother was just like that, she would stay there all day long and never get tired of it, never sit down, she liked to cook so it worked out well.

When I first came to Smith Island, Char was in her seventies, although she still owned and ran the island restaurant / coffee house

CHAR AND RUKE DIZE

called Ruke's. It was a place where local people gathered for comfort food and coffee. Many girls got their first job there, supervised by older women who worked for Char into their later life. It was also a welcoming locale for outsiders: the tourists and family visitors who mingled with the island people. The exterior of Char's restaurant resembled an old, wooden, crumbling, but poetic cottage, sitting on the edge of marshland. Inside was what people often defined as a "magical" place, with the kitchen on the right, separated from the dining room by a cash register on a long, tall table. The tables in the open, large room were surrounded by shelves throughout the room, displaying her prized antique collection, which defined the unique charm of the place.

When her husband went to a nursing home on the mainland, Char closed the restaurant and moved the items that were most special to her to her house. There, the old photographs were surrounded by the antique objects she had collected over her lifetime, such as old oil lamps, glasses, jewelry, teapots, and furniture. Char lived with the pleasant memories these objects brought her. She said, with a smile, "I miss Ruke, my husband, so much. We had good life together." After a pause, she continued:

> It was one time here, when Ruke was resting on the sofa while Clarence was visiting, and from nothing I hear this loud sound . . . *bam!* and something hit the window. I had no idea what it was and so I went to the kitchen to check what was going on. In my kitchen sink there was a wild turkey. Ha ha, can you imagine? I was so scared and didn't know what to do, I run out for help. Clarence left, and Ruke had no idea; he was left there alone on the sofa. Then I got Allan to come to get that turkey out of my kitchen.

After I heard Char tell this story multiple times to her friends, I asked her if this was now her "yarn," and she said, "Yes, that is." I was curious if women's storytelling, or telling women's "yarns" (a Smith Island term), resonated with storytelling elsewhere on the Bay. Her sense of humor, often expressed during my visits in her island home, became a part of my own experiences on Smith Island. Char's turkey story and her witty personality offered me a taste of the islanders' sense of humor, embodied in friendship.

Smith Island humor is manifested through the pranks, performed in the course of ordinary times, and playful joke performances and satirical skits that take place during many annual and life-cycle events, including funerals. The islanders' "joke work," which I observed during social events, is usually well-defined playful storytelling, such as Jennings's narratives about Smith Islanders' modes of knowledge that I discussed in chapter 1. Patty, who lived in Tylerton and was much younger than Jennings and Char, practices both: using joke work in daily pranks, and composing skits for social events. Patty's joke work is inspired by her social interactions, and she systematically takes notes when she is in the company of others. During family gatherings, she often carries what she calls her "joke bag," containing her collection of notes on loose papers, which are her records of what she finds funny. "I tell my family that I am taking notes, for them to know I am recording their joking and storytelling," Patty informed me. When I asked Patty about her interest in humor, she replied, "I think it was born in me. My grandmother, uncles, dad, and mom were all humorous of time. I enjoy things that are funny, and I am always looking for something that is funny that I can share. I like making people laugh and can make something funny about most anything. I work best around my home family."

For performances at collective events, Patty usually writes skits for each character: "I just look and listen and wait to see what the people around me say and always have a pen and notepad ready, and then I work it into skits." She writes down ideas that sound good to her year-round. Other island women also share their ideas with her, and then she decides if those would work for a funny skit. Her sister, Carol Ann, also likes to do joke work among her own family members. "People sometimes go to great lengths to pull off a prank. Our pranking can be crude, calculated, but also in the stories we tell, truthful stories," Carol Ann's husband told me. "First of April I made some pranks on my wife, but then she made one to me, not to be outdone. She put in my water bottle the vodka we keep for cooking in the kitchen. I didn't know 'til taking sip in the morning with my pills," he recounted.

Jerry also spoke about an older prank in a story about Tylerton: "I remember one trick that happened in the big store in Tylerton. Someone found a loose nail on the weatherboard and tied a string around it. Well, if you wet the string with kerosene and pull the string between your fingers in the store, it sounded like whole building was falling in. All the people in the store come running

The Taste of Things and Comic Relief

out." Some pranks, however, were pointed and harsh, and I was interested to know if there is a gender difference in pranks and joke work. "Men's jokes are maybe harsher, but my wife's prank was in the men's category; it was harsh I can say truly," noted Carol Ann's husband. Then he remembered a humorous tale that connects to the island tradition of storytelling.

> My dad tells me of a story of a widower man with a lot of kids. A woman's husband died, and at the graveside the widower man leaned close to the woman who just buried her husband and said, "I want to talk to you after this is over." He had plans on marrying her and didn't want others to get her first. People of the island laugh at this, because of the eagerness of the man. It may be unusual, but that gets a laugh. When I was in college, I had a class called "Dead and Dying," where we talked about being aware of grief and giving people time to grieve. I told this story, expecting people to laugh at the eagerness of the man and as an example of what not to do, but the class reacted completely different. No one laughed. Stories by islanders are primed to be received as funny on the island."

From a sociolinguistic perspective, humorous storytelling on Smith Island is integrated into all communications and, therefore, is inseparable from their language, whether in what is spoken every day or just used during celebratory times. Language play is part of daily chats on the road, at work, or in phone conversations. Humorous play takes on multiple forms during all "socials," or annual traditional and life course–related events. In this chapter I frame my discussion around the annual event known as the Ladies Dinner and examine the manifestation of female gender roles on Smith Island. Exploring how life in a particular place governs a woman's expression of her identity, I offer insight into how women's knowledge shapes local heritage through their daily work, as well as in gendered play during this once-a-year special occasion. I argue that in the context of Smith Island's environment, the resilience and sustainability of this fishing community, which is facing changes in its socioecology, depend significantly on ritualized collective engagements, in this case one initiated by women.

Women's Parodies and Ritualized Laughter

It was early December, and I was arriving on Smith Island for the Ladies Dinner. When I stepped off the boat, the soft midday winter sun was warming the island, and a gentle breeze touched my face. Like other annual social events there, the Ladies Dinner, held on the first Thursday in December, is well planned. This celebration follows the structure of other social rituals on the island and is built on three main elements: faith, food, and satirical comedy. Social events are generally based on a seasonal rhythm and the structure of Christian Methodism, as adapted to life on Smith Island. In their own unique style, the islanders' ritualized events combine prayer, storytelling, reading, singing, joking, and communal meals. Many social events involve men and women, but women typically will organize the meetings, where they plan meals and divide the necessary labor (cooking, cleaning, and decorating) prior to an event.

The Ladies Dinner begins with a communal prayer, followed by a traditional meal and performances involving music, storytelling, drama, and satirical comedic skits. Most of the island women look forward to this event all year. Outsiders, like me, are invited personally by women from the community. "You are trusted now that you were invited," said one of my friends, who had moved to the island as an adult. She went on to say, "There was real anger amongst the island women after a journalist from the *Washington Post* came to this event and published an article, which [island] women didn't find acceptable. They found it disrespectful and ever since have been guarding more carefully who they let come in." When I asked others about this incident, several women agreed that the published article was indeed disrespectful and made a mockery of the island. All Smith Islanders, not just the women, are very protective of how their lives on the island are characterized by the media and the press. The incident with the *Washington Post* reminds us of the ethical implications of crossing boundaries by publishing such information. Smith Islanders have established a fine line between "bad press" and "good press." They are open to sharing their heritage with the public, but they also recognize knowledge they believe belongs only within the community. What follows is based on my observations as an attendee at the annual Ladies Dinners between 2014 and 2021.

At just before 6 PM, the island was already dark, and the air around the harbor was filled with the steady clink of gear on the boats, punctuated intermittently by the sound of seagulls. As I approached the church, I could see some cars and golf carts parked out front. I opened the basement door and entered a room vibrating with the festive voices of women gathered from all three of the island's villages: Tylerton, Ewell, and Rhodes Point. Ewell and Rhodes Point are connected by a long road crossing the marshland, while Tylerton, which used to be connected by land, now sits apart, separated by water channels that have developed from erosion and rising sea levels. While each community has its own church and graveyard, there is a main general store / restaurant, along with a post office, in both Tylerton and Ewell. Those living in Rhodes Point who want to use these facilities must drive to Ewell, which is the largest of the towns and the hub of Smith Island's social life.

Members of each community have a strong sense of their town's unique collective identity. In their life history narratives, Smith Island women have expressed how much they miss their home community after changing residences. They speak of lifelong adjustments they have had to endure after marrying and relocating.

Residence patterns are not firmly established, as sometimes a man will follow a woman's inheritance and move to her village, but it is mostly the wife who moves to her husband's hometown. I was interested in knowing what, specifically, women miss most about where they grew up. When I asked Kristen, who had moved from Rhodes Point to Tylerton after marrying her husband, she replied, "I think it means missing the home life and the way things used to be. Missing parents and working together, spending time together, and celebrating holidays and special occasions. This is what I miss."

The location of the ladies' annual dinner party rotates among the three villages. Women from the hosting village cook a meal, decorate the venue, and buy or make a small present for each woman who attends. For example, one year, women from Tylerton and Rhodes Point attending the dinner in Ewell were seated at well-appointed tables, complete with a wrapped present at each place setting. Guests spent time socializing, while most of the women hosting the event finished their dinner preparations in the kitchen. The food was typical of traditional Smith Island cuisine. The Ladies' Dinner always includes two types of meat: chicken and pork. Side dishes consist of mashed potatoes, dumplings, green beans, corn pudding, and, often, a crab dish and fried oysters. Sweet black tea and coffee are the beverages of choice, and the special Smith Island multilayered chocolate cake makes a fitting dessert.

Once the food was set out and the guests had passed through the buffet lines and returned to their seats, two of the women from Ewell opened the evening's celebration by prefacing the meal with a welcoming prayer. The Ladies Dinner, like many other social gatherings on Smith Island, begins and ends with a Christian prayer. Words in the opening prayer typically are more general: giving thanks for the gathering and the well-being of the community, as well as the world beyond Smith Island. The closing prayer, in addition to giving thanks for the good time shared by the guests, includes a short theological reflection related to Christmas.

After eating, the women from the host community cleared the tables while guests waited excitedly for the highlight of the event: the comic performances. It is customary for the two visiting groups to perform satirical skits and slapstick comedy, interspersed with parody songs. On this occasion, one of the most memorable skits was a comic act that involved cross-dressing. A group of women wearing men's clothes and fake beards stood in a semicircle on stage, each of them imitating her husband's voice. Holding on to paper boats used as props, they acted out a parody of their husbands' conversations. As the women simulated the men's weather talk, crab-catching conversations, and other small bits of chitchat, the audience, recognizing the attitudes and mannerisms of the men, laughed vigorously. As an outsider who was not familiar with the different husbands' personalities, I missed some of the details, but I can still recall the sense of empowerment generated by the women's laughter. I can also still remember the pain in my jaws from laughing so much. This first skit was followed by a satirical parody of a visit to the doctor's office. The women riffed on their bodies' physical changes during menopause. With great humor, they dramatized the sudden occurrence of hot flashes and how they unfold, along with the lingering irritation associated with this female burden. Publicly joking about their changing bodies through self-mocking parody affords Smith Island women an opportunity to foster gender solidarity through such shared experiences.

Throughout the dinner, the women sang and danced between skits, leading up to the finale of the evening, in which the actors presented a slapstick comedy version of Julia Child's cooking show. Satire involving food and cooking is a part of each program, as well as the topic of menopause, with women joking about their hot flashes, forgetful minds, overall physical changes, and low level of tolerance for their husbands' "bad" habits. Over time, their ritualized language play and comedy increasingly shifted toward new topics, with performances about life in a nursing home or aging bodies, depicting themselves as more vulnerable and dependent on others in later life.

"We do this simply for our own joy, to give a gift to ourselves, and nothing other. We enjoy a good time together," said Jennifer, the manager of the island's senior center, as she showed me her album of photographs from past dinners. Over their life course, women have assigned multiple meanings to the Ladies Dinner. For many Smith Island women, the Ladies Dinner represented their first event on the island as adults. Kristen expressed her memories: "My mother was there. I just graduated from high school, and Ladies Dinner was my first event as an adult. It was an exciting time." Although using a different ethnographic context, José Limón (1994) also focuses on language through play and performance.

Events like the Ladies Dinner underscore the importance of ritualized gender solidarity for a community in which the women traditionally had to work together, running things while the men were temporarily away. Historically, Smith Islanders have depended on gender mingling to secure greater sustainability. Both men and women share an understanding of the need to sometimes assume the other gender's role in creating sustainability. I noticed a deep mutual respect that is embodied in life-long, lasting marriages on the island. During our conversations, some of the men spoke openly about a deep respect for these partnerships. Women's performances during the Ladies Dinner show how they handle marital unions with their sense of humor. In the past, when watermen worked on the Bay for weeks at a time, women had to manage their families and the island community on their own. But now a lot has changed. While women still support the men through the work they do at home and in the community, today men also support the women and their economies. Some men participate in the island's commercial baking business, making the traditional Smith Island cake, which is then shipped to the mainland. Some also participate in the tourist economy, traditionally directed by women.

∼∼∼

Reflecting on that night's parodies, along with similar memories, like that of Tylerton's Mary Aida and her famous Smith Island cake-baking demonstrations, I was reminded of what Paul Stoller (1989, 1997) discusses about taste in his ethnographic narratives. Because the skills needed to make Smith Island's traditional meals and the "taste full" memories associated with them are part of the island's pool of inherited knowledge, food is culturally, socially, and personally important. Sharing the tastes and flavors of these Smith Island foods at meals with family and neighbors, particularly during special events and rituals, is a manifestation of the unique identity of island women, as well as part of the island's heritage. Jennifer's comment (quoted two paragraphs earlier) highlights the community's gift-giving principles and directly connects to the concept of "gift-giving rituals" Marcel Mauss (1950) examines as part of social cohesion and reciprocity.

In ritualized events, such as the Ladies Dinner, women's gender-based satire provides an opportunity for them to confront established marital structures in liminal time and space (Turner 1969, 1974, 1977). Play—the free flow of artistic exchanges—not only exposes the limitations of reality, but art and play also ultimately create paradox and fiction. While gender mingling related to labor is usually shrouded in silence, annual events, such as the Ladies Dinner or the Waterman's annual dinner, reveal how playful Smith Islanders can be in their gendered knowledge. When women participate in the Ladies Dinner, they intend it to be a positive collective time, and they express this playfully in their skits and songs.

The language play to which they are socialized during their lives is manifested in the skillful free flow of such creative exchanges.

In addressing the complex levels of performance witnessed at the Ladies Dinner, my analysis turns to Judith Butler's (1990:187–188) concept of double inversion, which she conceptualizes in her feminist theory of imitative gender: "In imitating gender, drag implicitly reveals the imitative structure of gender itself—as well as its contingency." The Ladies Dinner is an event that offers Smith Island women an opening and the space where they can playfully invent what Butler calls "performative subversive acts." For her, gender parody is about making a "fantasy of fantasy, a double inversion." Gender parody, as Butler conceptualizes it, does not assume an original identity that the parody then imitates. Rather, a performance imitating gender plays upon the distinction between the anatomy of the performers and the gender that is being parodied. The shared laughter at the Ladies Dinners provides release during this ritualized gender inversion, and these comic imitations afford insight into Smith Island's treatment of gender fantasy.

My analysis of the Ladies Dinner underscores how such an event depends on collective participation and raises questions related to the future of ritualized heritage, not only on Smith Island, but also beyond this community. In a sense, women's agency is renewed during these events and has a beneficial effect on the entire community. Since only Smith Island women are asked to participate, the gradual process of population decline on the island has affected the number of women directly involved in organizing, performing at, and cooking for the Ladies Dinner. By inviting female relatives from the mainland, Smith Island women extend their social cohesion beyond the island. Yet early December weather often prevents kinswomen living on the mainland from attending. The boat ride to the island depends on that evening's weather, which makes it difficult to plan ahead for those who are older or who work. So the Ladies Dinner is evolving in a new context. Kristen's three daughters, in their early adolescent years, are part of the island's last generation of people born there, which consists of only about ten children. My observations of the women's annual celebration provoke further questions about the future of such ritualized, verbal creative play on Smith Island. Each year performed acts demonstrate an agency of womanhood by showing a range of topics that address issues related to the ordinary tasks island women do and the social difficulties they must endure. These may include personal illness and a loss of kin through death, as well as reduced prosperity or a damaged home during natural disasters. These painful experiences not only are acknowledged, but also are deeply shared in the dinner's final prayer. Therefore, this tradition that island women have established can "unite [us] in sisterhood," as one of the women put it. Many believe the Ladies Dinners will contribute to the overall well-being of Smith Island's community in the years to come.

Anthropological scholarship has shown that rituals are aesthetic human expressions, culturally constructed and relevant to our better understanding of what we may call the ordinary, day-to-day way of being in the social world. Declining numbers of women in attendance at the Ladies Dinner, paired with the diminishing number of future participants, leaves Smith Island women with an increasing sense of uncertainty regarding this time-honored, gendered ritual. By examining the nature of women's knowledge, I probe women's roles in reinventing heritage and highlight women's agency in relation to the dynamic tensions between continuity and innovation during a time of social change. I recognize that Smith Island women apply their overall knowledge to both masculine and feminine types of work, regardless of the generally established boundaries found in traditional communities. Similar to the findings of other studies focused on gender roles in fisheries, I have learned that the labor of

Smith Island's women and their social solidarity free the men to go out crabbing, oystering, and fishing. Women's daily work is focused solely on caring for their families and the community, since Smith Island women do not fish with men on the water. I discovered patterns in the women's work, as well as in the gendered rituals through which they manifest their agency and find power in the collective. The Ladies Dinner is a form of creative social interaction, employing aspects of both everyday life and the participants' skills and imagination. Their performances provide a lens for examining the agency that Smith Island's women express via their humor, knowledge, and faith. In considering Judith Butler's (1990) analysis, one can see how these gender parodies reveal that such dramatizations are an imitative structure of gender itself.

Gendered Knowledge and Woman's Flow

The stories told about the sustainability and resilience of the Smith Island community, recounted by the islanders and by writers on social media, highlight how traditional work is associated with masculinity. Smith Island's heritage is often articulated through the watermen's work of crabbing and oystering on Chesapeake Bay. But womanhood, as lived day-to-day on the island, is invisible to the public. Nonetheless, when looking through old family photographs, Jennings said to me with a smile, "We couldn't do it without them." "Them," of course, refers to the island's women.

While the watermen engage in their traditional masculine roles, women, in contrast, have had to develop great adaptability in the day-to-day gender roles embodied in their work on the island. As is the case in many rural coastal communities that are challenged by uncertain socioecological conditions, Smith Islanders have developed an understanding of the necessary balance between flexible attitudes, which are evident in the strong improvisational skills they apply to economic pursuits, and their traditional local knowledge.

"Women can do anything around here," responded Christine when I asked her and her neighbor Janice what they thought I should emphasize when writing about Smith Island women. "Women here can fix the house if they need to as much as they can do anything," said Janice, reflecting the attitude Smith Island women have toward taking on many types of work usually viewed as masculine. "Women here on the island are like pioneer women. They do what needs to be done," added Christine. In response to Christine's comment, Janice recalled a particular example of women crossing gender boundaries: "She [Christine] is the one who can fix TVs really well, everybody knew about the skills women on the island have and what they can fix." From women's accounts, like those of Janice and Christine, that I recorded between 2018 and 2019, one can see how women often speak of their agency through narratives about their work. Historically, women have managed daily life on their own, while watermen left for work on the Chesapeake Bay. Smith Island women do not work on the water, but they apply themselves wherever necessary, in response to the needs of their families, regardless of the "cultural inscriptions of gender" (Butler 1990). When Sheila spoke about her work, she emphasized the flexibility discussed by Janice and Christine.

> I love being a Smith Islander. I can't imagine living anywhere else. But for Smith Island women, it is hard work in the summer. We pick crabs, but it is really good money. This summer, so far, is not very good for picking because it's not many crabs. I have been cooking for men, planting grass on our bayside, feeding people at the B&B, and I do home care for the lady. And, I forgot, I sent three chocolate cakes this morning. Lot of Smith Island women make cakes and sell them. I am also a substitute teacher at the school. There are plenty of little jobs here if people would do them. We may not be very rich in money, but we're rich in other

JANICE KITCHING

ways. God gives us what we need to survive. We have to work hard in the summer to survive in the winter.

Because work is one of the most important aspects of a Smith Islander's life, many women like Janice, Christine, and Sheila (and men, too), would define their island life based on the daily work they do for their families and the community. Experiential learning of traditional work on the island takes place during childhood, often practiced in the company of kin. A first job, for many women and men, is a memorable time, marking their transition to adulthood in the community. For boys, their first job is on the water. For girls, their first job is on the land, often in service to the island community. "My first job was cutting the grass for the community," remembered Carol Ann, now the wife of the local pastor. Cutting grass on Smith Island is important work, because careful land management prevents the incursion of invasive marshland. Women's involvement with land management—expressed by an island phrase, "keeping up with waterfront"—speaks to my argument in previous chapters, emphasizing that land is observed, cared for, and managed with traditional skills and knowledge. But land is also identified as a gift from God, an intimate space aligned with the sense of self the islanders develop in the process of becoming.

Diversity is a distinctive feature of the Smith Islanders' knowledge base in relation to their daily work. Watermen generally engage in crabbing and oystering, while women's work has historically been more fluid. Women on the island constantly reinvent skills across a wide spectrum of jobs to sustain the community at large, as opposed to men's work, which is defined locally as traditional masculine labor on the water. Women on Smith Island move fluidly between their feminine roles and what is commonly defined as masculine work when they need to "keep up" (their term for supporting life in their environment). This division of labor does not suggest that women wield increased power in the community or at home. Women sometimes joke casually that men are just crabbing and oystering on the water, and women do everything else. Women's engagement with the island's water economy has changed over time, which is another example of their flexibility.

Most Smith Island women pick crabmeat for their friends and family. Women have learned their crab-picking skills by observing the previous generation of women who worked in pick houses. When picking together, the women working in Tylerton's co-op tell stories and jokes and sing island songs. At some point in their life, most of the women also work at least part time, if not exclusively, in community service jobs. In addition, many women are increasingly engaged in the island's ever-growing tourist economy, especially those from the younger generation, who find diverse work in the post office, school, restaurants, B&Bs, library, and senior center.

Traditional crabbing and oystering jobs on Smith Island are seasonal, physically demanding, and economically uncertain. They

DORIS LEE

are driven by the climate, ecological systems in and around the Bay, and market demands of the fishing industry. As the winter months begin and crabbing gives way to oystering, changing their work modes requires watermen to have multiple skill sets, along with an extensive knowledge of ecology, weather, work regulations, and the market economy. During the second part of the twentieth century, people on Smith Island witnessed a rapid exodus of their fellow islanders, who left for work on the mainland. Those who stayed and continued to work on the water have had to reinvent what they did, in order to sustain their incomes.

Since work on the water has been declining, this last generation of older watermen and women, when compared with their parents, will now labor longer into their later years. Women continue to balance their multitasking between unpaid domestic work and wage-based jobs, while Smith Island men mainly continue their traditional work on the water. It is the women who have reinvented their roles on the island to address its growing tourist economy, in addition to picking and packing crabmeat, "fishing up" (tending to the crabs in the shanty), and other work that supports their water economy or provides important services to the community.

Much of a woman's daily work also involves cooking for their families, neighbors in need, and the steadily increasing number of tourists. A unifying feature of Smith Island cuisine is the famous multilayered Smith Island cake. Not only is it the signature dessert of all island women, but it is also part of their heritage. Moreover, it has been designated by the state legislature as Maryland's official cake. The cake has evolved from its original version—yellow cake with German-chocolate icing—into today's multiple varieties. In addition to chocolate, women will make cakes using coconut, fig preserves, peanut butter, and various fruits, such as bananas and strawberries. When I first visited the island, a small number of women baked cakes for sale at the local co-op bakery, which has since closed. Most women bake at home now and sell their cakes to friends and family on the mainland or to the tourists and local restaurants on the island. Learning to make a Smith Island cake is viewed as a rite of passage in the life of every island woman. In addition to this traditional cake, many women develop a personal signature dish—that is, a particular meal or dessert they become known for. Considering the significance of this connection between "the taste of things" (Paolisso et al. 2019; Stoller 1989) and feminine roles on the island, I will now further discuss women's cooking and baking, experienced collectively as gendered sensory knowledge.

Sensory knowledge has become a common element in my understanding of Smith Islanders' "ways of knowing," particularly during my summertime fieldwork. When I have stayed on the island for an extended period of time, I have experienced their ways of engaging and being with the weather. Early on in my fieldwork, I had the opportunity to join men like Eddie Boy daily as they worked the water. As I became more accepted as a member of the community,

I gained a greater awareness of their gender lines, which I consciously avoided crossing as a female anthropologist. The limitations of a woman's (and of my own) movements that are ingrained in the island's culture not only reveal gendered geographies, but have also led me to discover the nuanced nature of women's sensory knowledge on Smith Island. As I experienced their sensory ways of being on Smith Island, I became aware of what constituted "tasteful things," which were often based around childhood memories and rooted in their understanding of a collective heritage.

Tasteful Memories

Individual women, especially in their later years, are known for some specific "tastes"—the flavors of certain foods they prepare—that others find exceptionally good. For example, people have said, "Edwina makes the best pineapple pie." Char, an older woman, told me, "Mildred makes the best stewed Jimmies." Sharing these stories connects people, through memories, to the personalities of long-lost

PEGGY CORBIN WITH HER CHILDREN

relatives. These "taste full memories," a term I borrow from Paul Stoller (1989), are passed down from mothers and grandmothers, as well as learned from neighbors and close family friends. When recalling memories of her grandmother, Joan from Tylerton spoke about the taste of Christmas morning: "She, my grandmother, came to our house early in the morning and screamed, 'Get up, get up! It's Christmas morning!' Then she would make the best fried potatoes ever. I will always remember her that way." When Bobby, a tugboat captain, talked about his grandmother when showing me her old photo albums, he recalled his first "taste full" memories of her: "My grandmother used to work at the store down in Tylerton, and occasionally I would go and visit her. She would usually allow me to get a dill pickle and a Yoo-hoo [a chocolate-flavored beverage]. She was the kindest person."

"Taste full" memories, as experienced by many Smith Islanders like Bobby, are a significant part of how people on the island relate to the community in their specific town. Each village has its own distinct traditional dishes, related to the local harvest, with recipes recorded in local cookbooks. Smith Islanders identify with a certain taste or type of food—a particular cuisine, which often consists of seafood they catch from the Bay—along with what many would typically call "comfort foods." The knowledge islanders accrue from cooking daily meals, baking desserts, and making traditional holiday meals is part of their local sensory knowledge, affording them a bit of local prestige. It is through this prestige, based on their association with both inherited and invented tastes—their community legacy—that individual subjectivities develop and are embodied in the gender-specific sensory knowledge kept by Smith Island women. In reflecting back on my fieldwork, I related best to Smith Islanders from my sensory experiences there.

Although the islanders dedicate most of their time to work, they highly value communal rituals. Current changes in the island's socioecological conditions—that is, population decline and land erosion—create conditions of vulnerability. But despite social hardships and environmental difficulties, the women in this community continue to organize social events. It is evident that in a time of environmental crisis, Smith Islanders, especially the women, deliberately and collectively engage in organizing communal celebrations. Before turning to a discussion of ritual, however, it may be helpful to first lay out the gendered nature of productivity and island life itself, particularly the ways in which knowledge is gendered.* Having explored the pivotal role women play in the work they do, I argue that such female-only rituals illustrate how women reinvent their identity and increase their power in the community through their social solidarity. My analysis of the Ladies Dinner underscores how such an event depends on collective participation, thus raising questions related to the future of ritualized heritage not only on Smith Island, but beyond this community. In a sense, women's agency is renewed during this type of event and has a beneficial effect on the community at large. Since only Smith Island women are asked to participate (although they can invite special guests), the gradual process of population decline on Smith Island has affected the number of women directly involved in organizing, cooking, and performing at the Ladies Dinner. By inviting women relatives from the mainland, Smith Island women extend their social cohesion beyond the island. Yet

---

* Anthropologists historically have recognized distinctly male and female rites. For example, in his late-nineteenth-century ethnography of Australian Aboriginal peoples, Sir Baldwin Spencer described gender-based celebrations called Corroborees, as did Franz Boas in his Northwest Pacific project. While many twentieth century anthropologists discussed gendered ritual practices through the lenses of various theoretical perspectives—such as the functional theories pioneered by Bronisław Malinowski and A. R. Radcliffe-Brown; the structural analyses advanced by Max Gluckman, Victor Turner, and Claude Lévi-Strauss; and, later, the symbolic/interpretative approach of Clifford Geertz and Mary Douglas—it was not until the 1970s and 1980s that feminist anthropologists Sherry Ortner, Rayna Reiter, Karen Sacks, and, later, Lila Abu-Lughod addressed gendered agency.

unpredictable early December weather often prevents their kinswomen on the mainland from attending, so the Ladies Dinner is evolving in a new context.

In addressing women's knowledge on Smith Island, this chapter has combined three analytical models: a socioecological framework (G. Bateson 1972; Berkes 2015; Ingold 2000), feminist theory (Butler 1990), and anthropological theories about ritual and performance (Limón 1994; Turner 1977). I have emphasized how a woman's role is defined by sensory knowledge, by "the taste of things" (Stoller 1989), in their relationships with others. In my discussion of how women engage with symbols of masculinity when performing their gender roles, I expanded my analysis of gender performance in relation to Judith Butler's (1990) theory of gender parody. By following the structure of a particular gender-specific celebration, I not only have discussed solidarity in women's voices, but also offered an in-depth look at the satirical skits performed on stage, showing what Butler (1990) explores in her analysis of gendered parody as "a double figure," elicited through the technique of comic inversion. In my analysis of these comedic parodies presented, I discussed the playful nature of such engagements and the subversions of gender, which women use when they are confronted with perceived gender boundaries. I then explored the practices and experiences embodied in both the ordinary lives of Smith Islanders and the extraordinary times during social events and rituals.

Anthropological scholarship has proven that rituals are aesthetic human expressions, culturally constructed and relevant to our better understanding of what we may call the day-to-day way of being in the social world. The overall declining number of women attending the Ladies Dinner, paired with the diminishing number of future participants, leaves island women with an increasing sense of uncertainty regarding this time-honored, gendered ritual. In examining the nature of women's knowledge, this chapter has probed the role of women in reinventing heritage and demonstrated their agency in relation to the dynamic tensions between continuity and innovation during a time of social change. I recognized that Smith Island women apply their knowledge to both masculine and feminine types of work, regardless of the generally established boundaries found in many traditional communities. Like the findings in other studies focused on gender roles in fisheries (Frangoudes and Gerrard 2018, Gustavsson and Riley 2019), I saw that women's labor and their social solidarity free the men to engage in crabbing, oystering, and fishing, while women's daily work is focused solely on caring for their families and the community. I discovered patterns in the work and gendered rituals through which women manifest their agency and find power in the collective.

The Ladies Dinner is a form of creative social interaction, employing aspects of everyday life, as well as the participants' skills and imagination. Their performances provide a lens for examining the agency Smith Island's women express via their humor, knowledge, and faith. One can see how these gender parodies reveal, in light of Judith Butler's (1990) analysis, that such dramatizations are an imitative structure of gender itself. This chapter has shown how fluid gender modalities, embodied in feminine roles, can not only make women's agency clear, but also evolve into agencies that define gendered local knowledge and heritage, which then serve to empower women in such communities. Yet, when thinking about how a reinvention of gendered roles empowers agency, recreating a heritage that is at risk, it also raises a broad set of questions addressing nuanced processes in the changing socioecology of grassland coastal communities that are exposed to global climate change.

7

The Art of Creative Futures

STEVE DUNLAP

Magical Grounds

"In the evening, when I am coming home from working on the boats, I turn into my street, I see the sunset, and I think to myself, I live a dream. I live on the island, and I can fix boats for a living. I live what people dream of. On the mainland, I will be just another white guy. I was there in that world, and I left because I didn't like it. The fact that you can live like this, and that this place still exists, is magical," Steve recounted.

I met Steve in 2014, shortly after I arrived for an extended summer stay on Smith Island. When settling into my rented house, I found out that the refrigerator was not working. Ken and Iris suggested that I ask their neighbor Steve to repair it. Ken walked over with me to make an introduction. Steve was outside in his yard, working on a car engine—one of his many repairs for islanders. Later that same day, he came to repair the fridge, and our conversation led to an invitation to use his small garden pool. Although Steve was usually busy fixing broken cars, bicycles, golf carts, boats, and household appliances, he and his partner Shawn still found time to befriend me and my family. Since then, our friendship has evolved, and spending time with Steve and Shawn has become part of what I do on each Smith Island visit.

When Steve became a boat captain, I traveled with him whenever I was leaving the island. During those boat rides, Steve talked about his life's journey and his Smith Island life. "It all came to one small story in 1994," he said, and then began to tell me about his pre-island life, when he was driving a truck for a living. He met Jemma when he worked for a trucking company. They began dating and made cross-country hauls together. One weekend, Steve went to stay with Jemma in her home in Annapolis, and she invited him to come with her for a short visit with her friends Barb and Virgil, a lesbian couple living on Smith Island. "It all comes down to this trip to see Virgil on the island. She was moving out, she couldn't make it work on the island, and we were helping her with her move," explained Steve. "We went for lunch to Ruke's Restaurant when I got this feeling about the island. Jemma and I just began to look for a house that we can stay in when we are not driving. I asked Virgil if there are any houses for sale on the island. She said, 'There is one just now and let's go see it.' That house became our house."

Steve and Jemma's Smith Island house "had nothing," as he put it. It needed new plumbing, electrical, and sewage systems. "It was a unique investment and I learned to do all the work on the house for the first time," explained Steve. The couple still traveled frequently, but the island house became their new base. He and Jemma married and continued to drive trucks for a living during the winter, and in the summer they stayed on Smith Island. In 1995, Jemma stayed in the house alone over the winter, while Steve was on the road. "That was hard. I couldn't always get there when I wanted. One time I got stuck there because of the weather and had to stay longer that winter," Steve remembered. In 1996 he put a new roof on the house. In 1997 they stopped long-haul driving and Steve started doing mechanical work for the islanders.

Jemma and Steve also worked for Char at her restaurant, with Steve at the cash register and Jemma waitressing. I was interested to know more about his first years on the island, and when I asked, he complied. "At first, we only knew some older people like Ruke and Char and Jennings and Edwina. Over time people realized I was trustworthy and reliable, and people accepted me for my skills. As much I was trying to fit in, I also just in some way fell into it, as I was not making trouble. I was also easy to accept," Steve reflected. "There is definitely a pattern of becoming a Smith Islander when a new person arrives," he added, in order to explain how newcomers may feel when they move to the island and desire to become part of the community. "Hardest part, for me, was to accept a small-town living, at first." Steve then described what his early years on the island felt like: "Nobody has secrets and people are friendly, but that doesn't mean that they are friends. I was from a big city where nobody knows you. When I realized this, I dropped my expectations, and it became easier."

When Steve separated from Jemma and came out as gay, his life dynamic changed: "I had to accept to be here as being different and after it became easy. I have been here twenty-five years and when I walk into the store, I can feel I am an outsider. Yet I also believe that if something would happen to me, people here would care." Steve emphasized how people trust him and value his services, yet he experiences limitations when he tries to make closer friends. "I hit a glass wall with islanders. I don't think they really care about gay things. They appreciate me for fixing their houses and that I don't rip them off. I am trusted to go to people's houses to do repairs when the house is empty or the wife is home alone, but in the beginning, when I try to, I organize parties, nobody came. So now I rely on my friends outside, and internet made it easy," he said, voicing his sense of alienation.

Steve joined the fire department, trained as an EMT, got a captain's license, and purchased his own boat. "I like to be useful in times of crisis and be available to others, visitors and islanders, for public boat rides between the island and the mainland," he explained. Acquiring a boat represented an important shift in his relationship with Smith Islanders. It gives him independence from the ferries, as he can come and go when he pleases. But also, in his view, "it gives me credit and respect, I think, because I am doing something like an islander, and I handle it well. I am different from people here. I am not religious, I don't watch sports or fish, but I am respected. I can fix anything and drive a boat."

During our conversations, Steve sometimes imagined that he would leave and stop "fixing the island," as he referred to his mechanical repair work. "If I left, they would be sad, but not reject me," Steve concluded. During our conversations about what it meant for him to live on Smith Island, Steve often spoke about his experiences in sensory terms. He emphasized how he was aware of the quiet environment, and how he liked hearing waves crashing from the Bay side and a horn sounding from far out on the water. "I pay attention to the stars at night, and the stars' constellations. I pay attention to weather, nature, and water," he stated when describing his sensory experiences. "Weather dictates everything I do. When I wake up in the morning, I look at trees and see if it is windy. I am not doing anything on the water if it is windy, foggy, or rainy. When it is too cold or too hot. I am controlled by the weather. Then I can stay at home in my PJs and rest."

Steve, like other newcomers, reinvented his way of life in the process of making a new home on Smith Island. He has to submit to the weather, a force all inhabitants must accept. Steve's unique stature on the island is defined by his fixing both broken things and ailing bodies, as his frequent involvement in health-related services on the island is part of his life. For islanders, material things are living organisms (with souls), which they themselves also possess. Steve's creative skills in fixing broken things inside houses, in shanties, or on cars and boats—in addition to saving lives as a paramedic—represent more than jobs he takes on in order to survive. It is his way of being on this island. It is through his caring for the brokenness in things and people that Steve's social identity is most recognized by islanders. Steve's story also shows how moving to Smith Island requires newcomers to keep an open mind and a flexible attitude during the process of assimilation. Establishing one's belonging within Smith Island's socioecology demands that newcomers reinvent their personhood and redefine their sense of self in relation to the island. While such experience can be "magical" in terms of the environment, we also hear conflicted "multiple registers" in Steve's voice (Bakhtin 1981).

Along with his different political views, Steve's ending his heterosexual marriage and openly engaging in a homosexual partnership made him feel suddenly alienated from his neighbors. But these newly realized changes in his social life did not force him to move away. Instead, he now understands that he must settle for a limited number of available friendships on the island and broaden his network of longtime friends outside the island community. He engages in many activities that contribute to community life on the island. He continues his repair work and his service with the Ewell Fire Department, and he operates a "summer-evening bar," serving nonalcoholic drinks, as the island community is dry. He offers boat transportation. He also serves as a DJ for the island's dance socials: "I am honored to organize dances. I am happy to see people coming and dancing." When I reminded him of this new tradition he had established, Steve corrected me: "Oh no, dances always existed, people got together for dancing before my time, but the rules of these social dance gatherings changed. Saturday night events used to always stop before midnight, but I keep playing music 'til 2 or 3 AM because people like it and stay." Dance socials take place in the fire department building, in the room above the fire engines, and the participants are part of a smaller group of younger and middle-aged islanders. Because most of the established social events on the island evolved into heritage practices over time, organizing dances has become Steve's way of participating in a creative reinventing of the community's collective heritage.

Steve's sense of belonging is complicated and can be painful for him. Shawn, Steve's partner, jokingly remarked, "We newcomers are misfits and old souls. When the boat is leaving the mainland in Crisfield, it is a magical moment. As the town is getting smaller and smaller, I feel like the energy is getting lifted." For Shawn, as for Steve, an island is a special place, one that brings him closer to nature, gives him joy, and makes him feel magical and alive: "When you live here, you live with the environment, you live closer to the earth. By touching grass and listening to birds, you remember childhood. There is beauty in the light and a sense of peace, and no bridge. You must put in an effort to come here, but when you live here day to day you sense how authentic and unique it is." Then he repeated, "It takes an effort to get here, a boat ride across the water, separation, but it is such relief. It is just a small town that happens to be on the water, but it has magical qualities." In contrast to Steve, Shawn has a more pragmatic view of the integration process.

> The island has a way of expurgating the wrong arrivals, the ones that try to change its identity. The ones that slide in fit well, who don't try to change things too much. I slid in. I was friendly to everyone, and I embraced (easily) the water world and culture, not really changing my own identity, but not trying to change the island at all, too. For one thing, there was no reason to change the island. The only thing I could maybe think to affect would be to show them a good example of a homosexual, perhaps, but honestly, it's not that big an issue here. You can be trans, you can be black or Jewish, can be liberal or conservative, you can be whatever. JUST DON'T BE LOUD. Just don't try to change the flow. Islanders find identity, comfort, and a sense of purpose in the routine, in resisting change. But there's the rub. In the face of nature, they are being forced to do something they so resist—change. Christmas play, when the actors now almost outnumber the audience, is symbolic, a testament to islanders' drive to keep their traditions alive and their very identity.

Steve and Shawn call Smith Island their home, and they are anxious to return here anytime they travel. While Shawn finds the islanders resistant to changes in general, he also noted that shifting natural, environmental, and energy patterns force the island's people to adjust and make changes in their own practices. Shawn remembered when Mark, one of the older islanders, once said, "The island just must face the fact that it needs newcomers. Without newcomers, the island is just going to die, and I'd rather have it continue for as long as possible, even if it means changing some." Shawn then continued our conversation by comparing his present life with his previous residency on the mainland.

> When I lived on the mainland, I never knew my neighbors, while here people like me, and when I need to connect with people, I can just walk out of my house and wait 'til somebody goes by and talk to them. That would never be on the mainland. Here you can engage with weather, water, and people. People here are kind when you live here, they are surviving together, and that takes precedence here. With the influx of new blood, the traditions and the way of being in the water are assuredly going to change more and more. It's a fine line between keeping their identity intact, their age-old ways of living, and their actual survival.

Social relationships present Steve and Shawn with many obstacles to overcome. Yet they both agree that, although it can be a real struggle, one must rise above ideological differences and relate to Smith Islanders as being who they are. Steve and Shawn view the island's people with compassion. They, along with the islanders, hold on to their hopes for Smith Island's future as they try to imagine what later life on the island will offer them. For newcomers like Steve and Shawn, the island is a different place today than how they imagine it was in the past. Smith Island is a place with new meanings and opportunities, but also one with many losses, due to its changing socioecology and climate. "The water is reclaiming the island, along with its culture. Like Atlantis. You used to be able to walk to Tylerton. Mother Nature has already started changing the map," said Shawn. When I asked Steve about what he meant when using the terms "magic" or "magical," he explained:

When I see what is happening with my yard, flooding, I think . . . it is a feeling of sadness over the loss that we then must endure, and how do we go about hope and making life then? When you come here and settle into island life, you become attached to the community. You love living here. Magic to me is in that you can witness things just as they appear. You can exist in space where you can co-create your own version of reality. This place is a bubble that can be whatever the person in it wants. Lack of what is normal makes it different from the rest of the common United States.

Steve and Shawn's world views are far from belief in a supernatural power. Their particular "magical experiences" on Smith Island lead us to think about modalities in which people formulate their relationships to a place. I like to consider two modalities of such magical participation: poetic magic and collective magic. I take poetic magic, in this context, to be a subjective intertwining of knowledge, imagination, and sensory experience that is related to

SHAWN McCREARY

an aesthetic perception of what a beautiful lived moment is in a place (Scarry 1999). I trace collective magic, referring to Steve's participation in the community, as a "magical consciousness" (Greenwood 2005). In her study of magical processes, Susan Greenwood theorizes about magical consciousness in reference to Gregory Bateson's (2000) description of magic as a process of thinking that is integral to the ecology of the mind. Greenwood argues that magical consciousness creates emotional meaning and is a form of associative thinking, bringing together interrelated sensory patterns.

Gregory Bateson (2000) discusses metapatterns linking social and planetary systems when he explores how the mind uses metaphors, symbols, and connections. While in the process of thinking, humans employ mental maps that link ecosystems and the mind through the language of magical experiences (G. Bateson 2000:467). In this sense, I argue that Steve and Shawn's participation in magical experiences is central to their imagining and way of making a life in this place. It is part of their way of being on Smith Island. I suggest that Steve's "fixing the island," to use his words, is a form of participation in Smith Island's collective magic. Considering how important Steve's repairs of broken items are for the islanders' well-being, I suggest that his work embodies a curative property, which leads us to think of him as a doctor for broken things. This brings us close to the concept of what Susan Greenwood (2009:54) calls "sympathetic magic."

Focusing on Steve and Shawn's island life reveals the conflicted realities some newcomers may encounter in the process of making a life on Smith Island. We can see from the two men's narratives how poetic, magical experiences are available in the vernacular landscape of the island's weather world. Yet we see, too, how imagined collective magic can turn into bad magic through a perceived alienation that is embodied in the collective. Reflecting on their experiences led Steve and Shawn to define their lived reality as a dichotomy. For them, Smith Island is a magical place in its environmental beauty and in peoples' care for others, yet it also presents a challenge in the feeling of alienation that arises when one's sense of social belonging fails. What makes Steve and Shawn's integration successful is their coming to terms with this dichotomous and conflicted reality. Their political views and lifestyle are different from those of other Smith Islanders, yet the island's socioecology holds magic for this pair. In the next section, I will turn to other newcomers and investigate their process of placemaking after they moved to Smith Island.

Injured Birds in Aram's Soul and Magical Things

Empty blue glass bottles on the branches of a small dead tree, old rusty engines given a bird's personality with stick-on eyes, a propane torch transformed into a plague doctor, and a rusty circle with a blue eye at its center are just some of the garden sculptures Aram Polster made from eroding objects found on the island. His art showcases

ARAM POLSTER'S GARDEN

ordinary objects in the light of a poetic magic found in materiality. Seeing Aram's art, so imaginative in its personification of material objects, I recalled the works of Surrealists like René Magritte, André Breton, Salvador Dalí, and Man Ray. They gave agency to peculiar found things, either through visual media, by blending the human body with open landscapes and architectural interiors, or, for Breton, through literary imagination. Aram transforms the objects he finds on the island into soulful sculptures. "Everything has spirit. Old rusty engines have an old soul," explained Aram in our conversation about magical realism in his art. "I can be creative here in this environment, more than in any other place. I like being near the water and collecting objects transformed by the environment and making them into birds. I am into metal. I am interested in how metal objects are changed when rusting. I find here things that nobody here understands why I pick up. They don't get it. I like the sight of decaying metal," he told me as he explained his art and his feelings about the island.

ARAM POLSTER AND HIS ART

Aram also makes lamps, often in birdlike shapes, from his sculptures, and many of these lamps illuminate the interior of his house. "Everything has spirits, animism," replied Aram when I asked him about his indoor-sculpture lamps. His house is a lived-in art gallery, transformed by the light from his art and the sounds of the birds outside. Significantly, Aram built a sanctuary for injured birds in his backyard, facing the marshland. The roosters, turkeys, and visiting seagulls who find comfort on his land are his family. Only the sliding glass doors separate them from Aram's kitchen. Sitting in the room he shares with Freddy, his large white parrot, Aram told me, "All the birds living here are part of my family. They have personality, and they are doing who they are. I love their wildness."

When Aram bought his house on the island, he was a she—Anne. Anne moved to Smith Island from New York City after graduating from law school, subsequently left to undergo a gender transition, and then returned to the island as Aram. "I am never trying to fit in. I don't really need people's approval, but any place you live, if you are a good person, you participate in and care for your community and the collective good, and people see it," he answered when I asked him about the islanders' response to his transgendered identity. His political and aesthetic views are radically different from those of most Smith Islanders, but despite his different way of living and seeing life, he relates to the islanders with compassion and serves his new community. By finding eroding metal objects and turning these into shining, imaginative sculptures, Aram transforms lost, meaningless objects into the concrete magic of his art.

Aram lives on the edge of the marshland, close to the village of Ewell. When I visited him in 2022, he reflected on the previous few weeks, when water surrounded his house during an unusual high tide on the island: "This time water come up into my house. If it were not for Steve, who repaired the damage, I don't know how I would have managed. I don't know. That was really bad this time. It really changes something for me, this time." I could hear from his distressed voice that this disruption of his dwelling place by flooding had been hard for him. It was not at all part of what he anticipated when he imagined his future on the island. That spring, floodwaters had reached many homes on the island and caused serious damage. Seeing images and comments on Facebook during the flooding, however, was not the same as being with Aram, in his house, and in close contact with his feelings and the swirling ideas in his mind about the rising waters. Frequent high tides change peoples' lives in multiple ways. As I mentioned in chapter 1, they prevent people from moving freely across the island. When the water rises to their doors many are trapped in their houses until it returns to its lower levels. Water's intrusion into their homes constitutes a permanent threat to islanders' lives. The soil in their gardens has become permanently soft and squishy. Its fertility is destroyed by the encroaching line of ocean water, killing trees and slowly transforming their green crowns into bare trunks. Ghost trees, many in birdlike shapes, poke up awkwardly out of the rippling grass that sways as wind brushes across the marshland.

Across the road from his residence, Aram bought a second historical house, where he established a coffee-roasting business and a B&B. Roasting coffee is a trade Aram began when he lived in Vermont, before his return to Smith Island. Many islanders and visitors enjoy his coffee, which is sold in local museums and a bakery. Aram's B&B is located next to the roasting room and office in his second house. "Most of the people who stay in my B&B love the old soul of my house, they enjoy it, and some come back," he told me.

Aram, like Steve, participates in what I call collective magic. In addition to his art and his regular job as an assistant on the school boat, he bakes for people and, at need, serves as a firefighter. Aram was also a leading force in establishing a dental clinic on the island. He wrote a grant to secure funding, contacted dentists, and developed a plan for a freestanding nonprofit dental clinic: "First year was touch and go, we had a lot of repairs in the clinic. Last year was the first year when dentists come over and stayed for two days. They are so nice and compassionate, and people recognize it." Aram emphasized the collaborative efforts to run the dental clinic on Smith Island: "Everybody helps. The dentists who come here are part of an organization called the Mission of Mercy; they come to do work for free. Women cook for them, and the clinic house has a space where they stay for two days. I was worried that nobody would come, but people here take advantage of this opportunity and come, even from Tylerton. People come together to help; some cook and help in dentistry, and others paid for equipment." The dental clinic now depends on funding from Smith Island United and is subsidized by the island's church.

Aram has a sense of integration into Smith Island's collective life through his engagement with the community. Yet he also has a strong sense of independence from other island people and desires to be alone. When I asked him about his plans to stay on the island, Aram said, "Nobody knows what will happen. For now, I am here, and I am involved as best I can, but I don't really need people, I am fine being alone." He has established some friendships and working relationships, but he also finds satisfaction in his solitude. Moreover, he has a deep bond with the injured birds in his sanctuary: "I love their wildness. They are themselves, embodied to my home. They have their personalities, they have a right to their language, they are who they are. I know them and they know our established relationship. They are my calling. They are where my money is going. My religion. My life. They are my family."

Although in a different manner than Steve, Aram also sustains his way of being on Smith Island through his participation in the island's magic. For Aram, magical reality is when he takes part in collective magic by engaging with the community, and when he is in the process of making magical objects. Even when both Aram and Steve emphasize that they are socially independent from the island community and are "doing my thing," we can see how their participation in a collective life is not only part of the process of placemaking, but can also be transformed into a collective magic.

In the next section I will continue my discussion of collective magic by returning to Kathy's life journey. Kathy has suffered from PTSD after returning from the Gulf War (see chapter 5). In her search for a peaceful place, she found Smith Island. Here, her self-healing practice has depended on her sensory engagement with the natural environment and her participation in the island's community.

Kathy: The Healing Grounds and Collective Magic

"I love being around these women, [they] inspire me and I like to give back to them. They never once make me feel like I am an outsider and I was so grateful for that," Kathy said when talking about the collective quilting parties she organizes for island women. "I must be creative. I have always done sewing for my kids when they were little. I also have my service dog now, but sewing helps people. Like any [kind of] creativity, like painting or writing journals, it helps with your moods." In chapter 5, I related a conversation with Kathy about her coming to Smith Island and healing in place. Here I address how, by organizing collective quilt making with Smith Island women, she not only revived the island tradition of quilting parties, but also established a new opportunity for herself, as a newcomer, to create deeper social bonds.

Kathy originally is from Princess Anne, on Maryland's Eastern Shore. She first came to Smith Island to visit her mother, who had already moved here. During that visit, Kathy bought an island house. Then, when she retired from the army, Kathy and her husband moved from Texas to Smith Island. I first talked with her after she had opened a quilting workshop in an old bakery building. She set up her quilting tables at the back of the store, leaving space at the front for an arts and crafts shop, where she could sell other locals' works, in addition to her own. "I do everything from old-fashioned to art quilts," she commented when I was looking at her work. Her quilting designs range from patterned symbols of traditional work on the water, to seascape ecology, to patriotic symbols, like an eagle, the Maryland State flag, and the American flag. She was enthusiastic about her plan for engaging the Smith Island community by organizing social events where people would do craft projects together. Kathy was also excited about making art with others on Smith Island. She opened her workshop to the island's tourists and organized a series of workshops for the residents. During the holiday seasons she began inviting island women to her studio for quilting workshops.

When I returned in the fall of 2019, I asked Kathy if she would let me film our conversations in her sewing room. She now held the quilting school in her home, since a new owner had purchased the building that housed her arts and crafts shop, and Kathy lost her lease. She calmly remembered what it had been like to be in her shop.

> When I started the store, I would have tourists coming in, and they were interested in local craftwork, so I sold local crafts. Then the store was bought, and I chose to step out. I miss the store. I miss people who stop by every day for a chat, for five minutes or an hour. Tourists ask funny questions about life here, like where we take water or whether we have electricity. Milty came every day, at the end of the day, when his sister, who works at the post office, was getting out of work. I can't express how much I am in love with this island. I love it. I came up with twenty-nine people on the island who make art, along with

their other jobs. Some, older, make boat models, and people paint and do photography and paintings on crab pots. They do it to keep busy and relieve stress and fight depression.

During our filming, Kathy described the quilt workshops.

> Last year we had classes here on the island. Only one woman could sew and the rest I had to teach when we started, but they all chose the most difficult quilt. When I first offered the class, nobody wanted to take it; they were so terrified. My philosophy is that I'd rather have it done, and fully finish the quilt, rather than have a perfect outcome. But then they did it, and they started to talk. Women used to work on quilts together and tell stories, and children would often play under the quilting table. Each woman started her quilt and kept a frame hanging under the ceiling [over the table] in the dining room. When others came to work with her, she would bring it down and they worked [on the quilt] together. That is how they did it.
>
> These ladies here on the island work so much all day, and they don't take time for themselves at all, so for them to steal a couple of hours per week and take time is what they need. They need it. You must take care of yourself before you can give. It's not just somebody coming in, sitting down, and paying for learning. Some made it into a perfect end product, but the whole point is that you play and never know what you will get.

While I was filming Kathy, I could hear the excitement in her voice as she was describing the quilting workshop. She seemed well satisfied with it. "Somebody made a mistake and we laughed, or when I said something and it didn't sit well, they laugh, but we just made camaraderie during these classes you would not believe. Some would be early and some late, some stay longer than the others, but you can tell they all wanted to be there. We gained as much companionship as learning how to make a quilt," Kathy stated, with a twinkle in her eyes. "They each put different colors and made different quilts. Let me show you what the ladies did. They made a complicated and hard one to make." Reaching into the pile of folded quilts to show me ones made by women at the quilting workshops, she said, "There are so many choices."

Kathy emphasized that the participants "quilt and talk, remember and laugh, yet it is still work. An active way of resting while making a tangible work, an outcome they can see and touch or show others." She saw how, for the island women who work all the time, collective quilting opportunities allowed them to step away from their ordinary jobs. By engaging others in quilt making and sharing her skills with them, Kathy provided creative opportunities for Smith Island women to return to an old island tradition, known as a quilting party. This, in turn, helped her manage her own mental health. As was discussed in chapter 5, Smith Island, in its quiet and beauty, is a place of healing. Yet Kathy's creative integration into this community provides as much healing for her as the ever-changing view from her window and the sound of the seascape surrounding her house. She qualified that, however, by saying, "Look, not everybody feels this way. People who like to be part of larger social groups or are on the go will not fit well here, but we [outsiders] are generally accepted. People wave and talk to us. When people move here, they are accepted." Our conversation led to a discussion of the reasons why people move to Smith Island and the acceptance they generally find: "There are often underlying reasons for moving to the island. You may not be a mainstream person, normal when you move to the island in the Bay, but you are accepted. And you are also accepted when your life changes, like Aram, who was Anne first and now he is Aram, or Steve, who was married first to a woman and now he is gay, or when you grow old and can't go out. People still accept you, they will come to you, visit you, and that is the point."

For Kathy, as for other newcomers, Smith Island is a place where she is nourished by poetically "beautiful" natural sites. She told me how the views from her windows, when she looks up as she takes a break from sewing, are visually stimulating and even healing. For her and for other newcomers, the island is a place where she can pursue her creativity. By engaging with women and children from the community, she not only found her social identity in her transformation, but also discovered collective magic in the healing process of searching for her long-term well-being. Similar to Steve and Aram, Kathy has been stimulated by the island's environmental poetics and has participated in its collective magic through the social process of quilting.

Creative Grounds and New Rituals

I was walking along the road when Pamela, one of the newcomers, stopped her golf cart and called out, "I am planning the first Smith Island Art Festival. We have a boat to bring all artists for a day. You must come." A year later, on the last Saturday in May 2017, I traveled on a private ship with other artists from the mainland, heading to Smith Island for the festival. At the crossroads in the center of Ewell, we saw Pam at the community-built tables for the festival's artwork. Islanders, artists, and visitors from the mainland came to show their art works, crafts, photographs, and books. The island women sold baked goods, and some well-established newcomers, like Chris and his wife Shelly, sold beef- and pork-steak sandwiches and root beer. Doctor Rob, who owns a house on the island, and Mark, one of the watermen mentioned earlier in this book, made and sold crab cakes and fried oysters. People typically drink iced tea, coffee, or soda, because the island is "dry," and alcohol is never part of the social events.

Kathy displayed her quilts, Aram sold his Smith Island roasted coffee, some men exhibited their boat models, and Missy, along with several jewelry makers from the mainland, presented their creations. Various island women exhibited homemade candles and objects made from pieces of sea glass. Pauli, an artist and newcomer who was one of the early initiators of the festival, displayed her island landscape paintings. A few regional writers who used Smith Island or Eastern Shore as the setting for their novels displayed their published books. At its conclusion, people seemed enthusiastic about the festival, and Pam was pleased to see her idea successfully realized.

Pam and her husband John moved to Smith Island after he retired from the army. "My husband and I lived a military life. Our last station was on Maddalena Island, next to Sardinia. You can only get there by plane and then by boat. We lived there with our children, and we loved it," she said, in order to explain why they were interested in island life during their retirement years. After returning to their home in New York City, they searched the internet and saw a posting about a house for sale on Smith Island. As Pam recalled, "Ruke's Restaurant was still open and Char and Ruke treated us well. We ate crab cakes and had a great time." Pam and John rented a golf cart and traveled around the island. When the cart broke down, one of the islanders stopped, helped them fix it, and invited them to a communal dinner that night: "As soon as we came to the church dinner, Char called 'Newcomers! Newcomers are here!' She was so welcoming and included us in their table group. That evening we sat with her group, all old-timers like Jennings and Edwina and Char and her husband. During the dinner people were coming to us, telling us about their family houses for sale. We had a great time."

Pam further recalled, "That evening, before we went to bed, my husband said, 'Now all we need is a thunderstorm and it would be perfect. I can die here.' I said, 'I can live here.'" Pam and John returned several times to look at houses in all three villages, and they eventually bought a house in Ewell. "That was what did it.

People welcomed us right from the start. We loved the community and living on the island. We were accepted, and they held a welcoming dinner for us, I mean an older generation. It took a little longer with a younger group," Pam remembered. The couple got involved in communal life soon after they moved to the island. John did some administrative work for the community, and Pam worked at the local restaurant during the summer season. In 2015, Pam began organizing the first Smith Island Art Festival. When I talked with her a few years later, I asked her why she initiated a festival and how the community responded.

> When [Hurricane] Sandy hit and Smith Island United was formed, men from the island asked John to help. One of the future goals was to diversify the island's economy. Many voted for tourism. There was a musician visiting the island, and I asked him if he would play at the art festival. He agreed, and so I asked Pauli, an artist, if she would be involved, and she agreed. Then I asked Betty Jo if she would support the festival with her restaurant, and she said yes. I further asked Janet, from the cultural center, if we could use the porch for the festival, and she agreed. I invited Kathy to show her quilting. She first didn't think it was art, but she agreed. Donna and Maxine contacted me and asked if I would consider their cakemaking, and I was very excited about that. Sherrie raised money for a boat to bring artists for free from the mainland. We had 300 people coming to the island for our art festival.

Pam had been crafting her ceramics for a few years, and when she and John moved to the island, she made one of the rooms in their house into a ceramic studio. She sees herself as a creative person, an artist. While organizing the art festival was something new for her, this was her way of contributing to the island's future sustainability. She discussed with me how, in her view, organizing an art festival on Smith Island was integral to the island's future vision for developing a tourist economy. For her this was a logical fit: "When people come here, we need something for them to do, we need to engage them." Pam further explained, "We have watermen's tours in shanties, but also there are creative people on the island, like boat-model makers, or jewelry artists, and the decorative craft makers, and this is a good opportunity for them to show their work and connect with outsiders." Pam's initiation of the Smith Island Art Festival is another example of how newcomers engage with the islanders through their creativity, while also responding to their personal callings and the community's collective needs. Her lifelong interest in making art translated first into teaching ceramics classes for the islanders and, later, into organizing the Smith Island Art Festival. Seeing how well the islanders accepted this art festival shows how, in this community, with many traditions inherited from their ancestors, people are open to adopting new social events and integrating them into the island's traditions. This encourages us to view heritage as a dynamic force in the process of creating traditions that contribute to sustaining ways of living in existing communities.

The Smith Island Art Festival continued for five years, but local participation declined over time. During its sixth year, the majority of participants, both merchants and visitors, came from the mainland just for one day. When I asked about this low local participation, several people told me that the women selling their cakes the previous year had too many cakes left over, and the stand with fried oysters depended on an oyster farm that was not successful that season. It was as if momentum for this new tradition was gone when Pam began to travel and spend most of her time on the mainland with her new grandchildren.

Following Steve, Shawn, Aram, Kathy, and Pamela through their various processes of integration into the Smith Island community enabled me to offer insight into how this happens. The process of integration is always a subjective experience, as many newcomers have distinctly different backgrounds and motivations for moving to Smith Island. Their successful integration into island life depends on many factors, such as personality and economic background. These variations can be overcome if creativity and imagination are brought to bear on their personal and social lives. Thus they can address their own needs, as well as contribute to the sustainability of the island. By following Steve's journey, I have documented how seeing beauty in the landscape and providing service for others can bring a "magical" emotional experience and overall satisfaction to his island life. In exploring Aram's engagements with the island's environment, both artistically and socially, I have expanded on my early discussion about Smith Island as a creative ground. Kathy's narratives have shown how her healing process and artistic creativity have been manifested by organizing a quilting group, which not only has helped recover her health, but also revived a local heritage. Pam's position in the community is very different from Steve's or Aram's, yet, like them and like Kathy, she adjusted to life on the island when she established her identity as a part of the creative process. By inventing new economic opportunities for the advancement of tourism, and through her vision of the island's future, Pam instigated a new tradition, the Smith Island Art Festival, although it is uncertain if it will continue as an annual event. For many newcomers, as well as Smith Island old-timers, dealing with island heritage is always a question of adjustment, reinventing it into a new social practice. Otherwise, a ritual is lost when it no longer serves this island community.

New theoretical perspectives regarding community-based heritages that are impacted by changing ecology—which previously had rejected traditional views of heritage conservation—now argue that heritage, in its nexus with climate, is not a passive phenomenon that is destroyed by the climate's forces. Rather, it is a dynamic political process, depending on a complex web of local sustainability (see Harvey and Perry 2015). Historically, scientific interests in climate mostly addressed past climatic events, while the current climate-change discourse is oriented toward climate futures.

In this chapter, I have shown that a closer look at the adaptation and agency of newcomers as they establish their ways of life on Smith Island suggests that constructing heritage futures is a dynamic, ever-changing process, linking the past and the future. Subjectively, it is an expression of social values embodied in time and independent of environmental limits. My analysis here has shown how newcomers bring a creative force into a collective effort to reimagine the island's future. Despite real and perceived losses, both social and environmental, the remaining Smith Islanders, along with the newcomers, are moving forward in tandem. By becoming part of the fabric of current social life, these newcomers inject more energy and a new perspective into the process of redefining the island's heritage. They bring a creative force to the readjustment demanded by a changing socioecology. As old-timers and newcomers carry out their daily tasks together, they secure their joint survival. As northeasterly winds repetitively push water from the sea deeper into the island, flooding erodes both the land and newcomers' imaginations about future magical experiences. Rising waters not only flood the land, but also alter peoples' perceptions about aging in place. It is uncertain which newcomers will sustain their life on the island and which ones will be forced to leave, due to their circumstances. For Smith Island residents, dynamic population movement and chance bring a familiar uncertainty. Just like the water, which comes onto and runs off of their land, in many cases newcomers may come and then, as they say, "go off the island."

Epilogue
ETHNOGRAPHIC POETICS

Human trajectories that are written into places speak about people's engagements with their environment. People create diverse ways of belonging to the social landscapes they inhabit. This book has shown how those dwelling on a small island develop shared knowledge and deep emotional attachments to that place. I have discussed social, cognitive, and sensory modalities of knowledge while presenting life histories and my observations of collective life. I explored how Smith Islanders' sense of self is connected to their sense of place. Observing how people construct their world, and how the environment, in return, can shape their lives, I showed how they connect the material world with their values by using their skills and imagination. Describing the islanders' work, kinship connections, and social life, as well as their emotions, values, and personal poetics, I emphasized their way of life in their weather world. In my analysis, I looked at how individuals make sense of their conditions and examined how people's practices and beliefs are integrated into their local ecological systems.

Epilogue

After sharing the overall experiences of Smith Islanders, I addressed how aging islanders are responding to their changing socioecology as best they can—"going with the flow," as they put it. I emphasized how, for people like Smith Islanders, individual experiences and collective life are closely intertwined with the surrounding seascape. The ever-present uncertainty that islanders endure is caused by weather, life in the water, and a new uncertainty that is threatening Smith Island's future: the changing climate. Thus I have regarded the island itself as a character in the residents' narratives. As the land is changing, they are losing their island grounds to the water. Living trees are turning into ghost trees, no longer useful for people and birds. Friends and kin have moved away, and many others are dying from illness or old age. Living in such a paradigm is not easy, especially for people who, like most Smith Islanders, are in the later stages of life.

While writing this book, I found myself hesitant to take a positive or negative view of the Smith Islanders' current conditions and their future. I recognize that the islanders do feel a burden of sadness from multiple losses—of their land, their traditions, and their kin—but they still maintain a sense of hope that their efforts to adjust and redevelop their ways of living will continue to create new possibilities for life on what remains of their island. Frequent, damaging flooding clearly indicates rapid ecological change and forces us to recognize human limitations, even in what is a technologically sophisticated community. This book suggests that environmental pessimism is not a sustainable outlook for people in small island societies that are affected by global climate change. Therefore, pessimism should not be part of our way of thinking about the human dimensions of climate change.

My research has demonstrated the strong need of people in a place like Smith Island to hold on to their community and their subjective sense of self—of being in place. This ethnography lies between hope and decline, between lost identity in a vanishing locale and the powerful will of people to preserve their traditions as they imagine their island's future. Anthropologist Michael Jackson notes that such a double bind would be absurd if a person was alone, but since people live with and for other people, it is in their relationships with others that they find the strength to recover from and be reconciled with their deepest losses. This book has highlighted such recoveries. I have shown the resilience of the islanders in my multiple discussions about the residents' collective, everyday experiences on the island: at work on the water, at home alone, in the church, with the community, and walking or riding on the roads. Their daily lives show us that although they work hard every day, from dawn 'til night, they also value a sense of play and find joy in the company of others and in the poetic experiences of their surroundings.

Anthropologists are very well aware of how much our work and our long-term, in-depth engagements with local communities change our professional and personal views of other peoples' lives. We know, too, that our own way of life can be influenced by the communities we study. When we, as anthropologists, dwell with people, our own knowledge, emotions, aesthetics, and even ideologies are shaped by the individual emotional needs and collective experiences of those around us. Jennings showed me, through his narratives and photo archive, how humans' capacity for the poetics of social relations in place can transform a sense of uncertainty and helplessness embodied in the harshness or loss of island life. I saw how, despite his precarious, demanding physical work in the seascape weather world, the poetics of life became a healing process for him after losing his family.

While my affection for Smith Island's people and their place grew steadily during the years of my fieldwork, I also started to experience my own sense of place and community there. I formed friendships, and I had to learn how to endure multiple losses—not just of people, but also of the landscape. On my most recent trips there, I encountered the loss of land claimed by the water and a landscape scarred by dead trees. Even worse was the absence of conversations, meals, and prayers with my old friends.

Attending and helping at Smith Island funerals between 2015 and 2020 has changed my own understanding of human grief and, in particular, my view of the collective power of reconciliation with death. I noticed this change in my thinking when, with deep regret, I was unable to attend Iris's funeral in May 2020, just two months after the death of her husband Ken. His funeral marked my last island visit before the COVID-19 lockdown. Nonetheless, I was left with my memories of Iris. They mostly focused on her hands, visualizing them holding the chalkboard she had used during our last conversations. I realized then how my participation in Smith Island funeral celebrations changed my relationships with the island's community, as well as contributed to my new understanding of the grieving process. As I grew closer to island families, my participation in their funeral rites changed my position in the community. I was no longer a distant, silent observer. Instead, I had become an active participant in the lives of others on the island. I am now welcome to share my memories during services at the church, and I am part of the group that serves dinner at gatherings after the funeral.

Frustration from having been denied my opportunity to reconcile Iris's death in the company of others reminded me of Jennings' two poems, as well as Hoss's final gift-giving to deceased fellow islanders. Since I could not travel to Iris's funeral, I wrote a poem for her and asked the pastor to read it during the proceedings. The process of writing this poem gave me a sense of relief after losing my close friend. It was an expression of my emotions for Iris, and I consider it a final gift to her. Later, when another island friend, Char, died, I wrote a poem for her and read it at her funeral. After that, during a December evening on the island, Edwina asked me if I would write a poem for her when she died. She then told me that she had been newly diagnosed with cancer and had six months to live. I said yes, and I started to write it that night, but I was not able to finish her poem until after she had died. The loss of Iris, Char, and Edwina was a bitter reminder of the ever-changing stream of life. Even when I try reasoning with myself in advance about impending deaths of the oldest islanders, in the hope that I would convince myself to be prepared for their loss, I still feel that I can't ever be ready to easily let go. So, to ease my feelings for the women I grew close to during my fieldwork, I wrote the following poems:

THE LAST LETTER TO IRIS
I was a stranger to you
first in your boat and then in your island house
You let me watch you cooking through the collection of your
    glass
I was a stranger at your table
I was a stranger during your prayers
And when you shared with me the taste of things:
Crab cakes, oysters, cornbread and grits, bacon and eggs,
    layered cake, and sweet tea.

I hold on to your smile
your taste in food
your prayers
the light traveling through your glass collection
In my memories
for twenty years
until I saw you again
always elegant with makeup on

in your white floral robe
you were an early riser in the island mornings.

When I brought my children to you
you embraced them
like they were yours
you held their hands
you looked at them with that sparkle in your eyes
and smile.
You had stories and dreams to share,
and when you didn't like something
you always said it like it is.

I was by your side
at the church,
the watermen's dinner
and ladies' Christmas night.

When the time came to leave the island
uncertainty made your voice sound
like a bird in a storm
disconnected when flapping its wings in the wind
hoping to be reunited with Ken.

In those last years, you always greeted me with rejoicing eyes
as if you accepted that you would not live long to see me again
I was sitting by your nursing home bed
in the room that was always hot in its smallness
and you said, "I am ready for the Lord . . .
I am ready, but he wants me to stay"
lingering between the earth and sky
in memories of the sweet wind from marshland, you used to say,
"I would give everything to be back on the island."

So now that you made it to the boat and back to the island
You are where the sweet smell of marshland and blowing wind
    is, your home
I will always remember your hands holding the Bible, with the
    loss of its gilt-edged pages
your eyes were swept by surprise
your written words inscribed into the little blackboard after you
    lost your hearing
Iris standing there alone in the end, woman and God!
and like you said, Iris, "I Love you!"

CHARLOTTE, THE KEEPER OF BEAUTIFUL THINGS
Teapots, hurricane lamps, and old framed photographs,
Salt and pepper shakers, jewelry, the piano, and a royal sofa.
Char, for Charlotte, all these things you collected were safe in
    your places:
Your home and restaurant, but also your mother's house—a
    place with ocean views where you returned in your
    imagination.
Char, you celebrated the colors and designs of all things,
    through your sense of beauty,
You transformed soulless objects into things embodied in
    stories.
What others may not see in the materiality of objects,
And some may consider it only old junk,
Char, you found meaning and magic.
Many things came to life in your gaze; for me, this is magical
    realism.
Challenged by the hard-headed things, such as the water heater,
    your car and golf cart, TV, and AC, you always smiled in
    good humor waiting for someone to come fix them.

Things would become your soulmates in a time of long, lonely days.

Char, from the dawn of your life on the island, through the loss of Ruke you kept old stories, women's yarns, like one about the wild turkey that flew through the open window,
landing in your kitchen sink—one you used to tell over and over with humor.
Others kept your stories—like one Jennings tells when in the first grade you decided to leave school in midday because, as you said, "I am hungry, and I am going home!"
You follow your voice Char!
Just like one day when you told Ruke, "I will open the restaurant, whether you like it or not!"

You were the keeper of kindness for others when some people were lacking.
For strangers who came to your store or island, from near or far, and for life-long friends.
Your words were warm and witty,
whether it be ordinary chat or the print on a birthday card,
like the one you wrote for Mildred, she has treasured that amongst old family photographs.
Char, you knew about the power of words but didn't go for gossip.
Some memories never failed you, like the ones about working with your mother or those of your marriage and family, even if some left you with a sense of the incomplete.
Serving in your restaurant, you offered magic with the food.
Coming to the funerals of passing friends in your striking jewelry against a dark blue dress,
you always showed up for others, keeping your smile.

EDWINA EVANS

EDWINA

I knocked on your door
In early December 2021
I saw you in your solitude
sitting in the armchair Jennings, your husband, used to sit in.
The living room walls guarded your lifetime memories.
The joy and pain you endured.
You smiled when I walked in, but there was stillness in your eyes.

Time had stopped like the clock on the wall.
We sit in silence after you told me about cancer growing inside you.
Your eyes filled with tears as your hand touched the belly.
It was a time without words.
I held your hand and you asked,
"Will you write me a poem when I die, as you did for Char?"
"I will," I said.

That night I began writing a poem for you
with the hope that we will read it together.
But I could not finish it; I had no voice for a poetic closing to
     your life then.
Only now, in your peace, I find the voice to honor your life.
I honor your playful joy lived through the sisterhood, with
     Hester and Faye
I honor your hope in waiting for Jennings stationed with the
     army in an unknown faraway place.
I honor your motherhood, challenging you
by what others find unbearable—a mother outliving her child.
I honor how the pain you endured did not stop you from caring
     for Jennings.

From shared prayers, the laughter and storytelling over the old
     photographs
you made your home into a seemingly eternal shared place,
and later into habitual solitude after everybody left.

Jennings retired from writing poetry.
But he will always have a love poem for you in the corner of his
     eye
I am just filling in for him when I write this poem.

I, we, celebrate your beauty which Jennings never stopped seeing:
Wearing sky-blue or iris-purple sweaters,
caring for your hair with a smile,
the music of your voice that never left conversations
demanding of you a close reading of others' lips when your
     hearing was gone.

I honor how you endured the pain with grace
and kept a playful sparkle in your eyes,
even after all those trials.

Acknowledgments

This book is the result of the tremendous support and friendship of many people on Maryland's Smith Island. I am especially grateful for the hospitality of island residents Jennings and Edwina Evans, Edward and June Evans, Elmer Jr. and Mary Ruth Evans, Peggy Evans, Faye and Wesley Bradshaw, Mitzi Brimer, Eddie and Joan Corbin, Eddie Sommers, Pastor Jim Evans and Pastor Everett Landon, Maxine Landon, Mary Aida and Dwight Marshall, and Jerry Smith. Newer members of the Smith Island community—Steve Dunlap, Aram Polster, Pamela and John DelDuco, and Kathy Taylor-Donaway—have also been very helpful to me. I want to thank Johns Hopkins University Press for their interest in publishing my work, and Tiffany Gasbarrini for her enthusiasm and efforts in getting my book ready for publication. Many people have generously supported my work through many hours of proffering advice and editing. I especially thank Margaret Supik for the editorial labor she has devoted to my writing. Because of her dedication and perseverance, this book has textual clarity and accessibility for a wider audience, rather than being limited to an academic readership. Special thanks go to Dr. Geoffrey Burkhart for his endless readings of my manuscript, his intellectual inspiration, and his continued belief in my work. Thanks also to Dr. Janice Harper for her editorial work and overall advice. I want to thank Dr. Deborah Van Heekeren for her academic advice and final editing.

I would especially like to acknowledge the support of my family: a thank-you to Frank Rehak for his editing of the first draft of this book, and a special thank-you to my daughters, Frances and Ester Rehak, for their support, humor, and companionship during my fieldwork on Smith Island.

Bibliography

Abram, Simone, Jacqueline Waldren, and Donald V. L. Macleod. eds.
  1997. *Tourists and Tourism: Identifying with People and Places*. Oxford, UK: Berg.

Abu-Lughod, Lila.
  2016. *Veiled Sentiments: Honor and Poetry in a Bedouin Society*, 30th anniversary edition. Oakland: University of California Press.

Agar, Michael.
  1994. "The Intercultural Frame." *International Journal of Intercultural Relations*. 18 (2): 221–237.

Alley, Dawn, and Eileen Crimmins.
  2007. "The Demography of Aging and Work." In *Aging and Work in the 21st Century*, ed. Kenneth S. Shultz and Gary A. Adams. Pp. 7–23. New York: Routledge.

Allison, Anne
  2013. *Precarious Japan*. Durham, NC: Duke University Press.

Anderson, Eugene N.
  2011. "Drawing from Traditional and 'Indigenous' Socioecological Theories." In *Environmental Anthropology Today*, ed. Helen Kopnina and Eleanor Shoreman-Ouiment. Pp. 56–74. New York: Routledge.
  1996. *Ecologies of the Heart: Emotions, Belief, and the Environment*. New York: Oxford University Press.

Bakhtin, M[ikhail] M.
  1981. *The Dialogic Imagination: Four Essays*, ed. Michael Holquist, trans. Caryl Emerson and Michael Holquist. Austin: University of Texas Press.

Basso, Keith H.
  1996. *Wisdom Sits in Places: Landscape and Language among the Western Apache*. Albuquerque: University of New Mexico Press.

Bateson, Gregory.
  2005. *Angels Fear: Towards an Epistemology of the Sacred*. Cresskill NY: Hampton.
  2000. *Steps to an Ecology of Mind*, University of Chicago edition. Chicago: University of Chicago Press.
  1999. *Mind and Nature: A Necessary Unity*. New York: E. P. Dutton.
  1972. *Steps to an Ecology of Mind*. New York: Ballantine Books.

Bateson, Mary Catherine.
    2013. "Changes in the Life Course: Strengths and Stages." In *Transitions and Transformations: Cultural Perspective on Aging and the Life Course*, ed. Caitrin Lynch and Jason Danely. Pp. 21–34. New York: Berghahn.

Berkes, Fikret.
    2018. *Sacred Ecology*, 4th edition. New York: Routledge.
    2015. *Coast for People: Interdisciplinary Approaches to Coastal and Marine Resource Management*. New York: Routledge.

Berkes, Fikret, Johan Colding, and Carl Folke, eds.
    2003. *Navigating Social-Ecological Systems: Building Resilience for Complexity and Change*. New York: Cambridge University Press.

Boas, Franz.
    1964. *The Central Eskimo*. Lincoln: University of Nebraska Press.

Butler, Judith.
    1990. *Gender Trouble: Feminism and the Subversion of Identity*. New York: Routledge.

Carey, Mark.
    2010. *In the Shadow of Melting Glaciers: Climate Change and Andean Society*. New York: Oxford University Press.

Casey, Edward.
    1993. *Getting Back into Places: Towards a Renewed Understanding of the Place-World*. Bloomington: Indiana University Press.

Charles, Anthony Trevor.
    2001. *Sustainable Fishery Systems*. Fish and Aquatic Resources Series. Oxford, UK: Blackwell Science.

Chatterji, Roma, ed.
    2014. *Wording the World: Veena Das and Her Interlocutors*. New York: Fordham University Press.

Crate, Susan A.
    2021. *Once Upon the Permafrost: Knowing Culture and Climate Change in Siberia*. Tucson: University of Arizona Press.

Crate, Susan A., and Mark Nuttall, eds.
    2016. *Anthropology and Climate Change: From Actions to Transformations*. New York: Routledge.

Cronin, William B.
    2005. *The Disappearing Islands of the Chesapeake*. Baltimore: Johns Hopkins University Press.

Dalidowicz, Monica.
    2015. "Being Sita: Physical Affects in the North Indian Dance of Kathak." In *Phenomenology in Anthropology: A Sense of Perspective*, ed. Kalpana Ram and Christopher Houston. Pp. 90–113. Bloomington: Indiana University Press.

Danely, Jason, and Caitrin Lynch.
    2013. "Transitions and Transformations: Paradigms, Perspectives, and Possibilities." In *Transitions and Transformations: Cultural Perspective on Aging and the Life Course*, ed. Caitrin Lynch and Jason Danely. Pp. 3–20. New York: Berghahn.

Desjarlais, Robert.
   2015. "Seared with Reality: Phenomenology through Photography, in Nepal." In *Phenomenology in Anthropology: A Sense of Perspective*, ed. Kalpana Ram and Christopher Houston. Pp. 197–225. Bloomington: Indiana University Press.
Dize, Frances W.
   1990. *Smith Island, Chesapeake Bay*. Centerville, MD: Tidewater.
Dove, R. Michael.
   2014. *The Anthropology of Climate Change: An Historical Reader*. Hoboken, NJ: Wiley / Blackwell.
Evans-Pritchard, E. E.
   1940. *The Nuer: A Description of the Modes of Livelihood and Political Institutions of a Nilotic People*. New York: Oxford University Press
Edwards, Elizabeth.
   2001. *Raw Histories: Photographs, Anthropology, and Museums*. Oxford, UK: Berg.
Fabinyi, Michael.
   2007. "Illegal Fishing and Masculinity in the Philippines: A Look at the Calamianes Islands in Palawan." *Philippine Studies* 55 (4): 509–529.
Fiske, J. Shirley.
   2016. "'Climate Skepticism' Inside the Beltway and Across the Bay." In *Anthropology and Climate Change: From Actions to Transformations*. ed. Susan A. Crate and Mark Nuttall. Pp. 319–325. New York: Routledge.
Foley, Ronan, Robin Kearns, Thomas Kistemann, and Ben Wheeler.
   2019. *Blue Space, Health, and Wellbeing: Hydrophilia Unbounded*. London: Routledge.
Frangoudes, Katia, and Siri Gerrard.
   2018. "(En)Gendering Change in Small-Scale Fisheries and Fishing Communities in a Globalized World." *Maritime Studies* 17: 117–124. https://doi.org:10.1007/s40152-018-0113-9.
Geertz, Clifford.
   1983. *Local Knowledge: Further Essays in Interpretive Anthropology*. New York: Basic Books.
   1973. *The Interpretation of Culture: Selected Essays*. New York: Basic Books.
Gower, Jon.
   2001. *An Island Called Smith*. Llandysul, Wales: Gomer.
Greenwood, Susan.
   2020. *Developing Magical Consciousness: A Theoretical and Practical Guide for Expansion*. London: Routledge.
   2009. *The Anthropology of Magic*. Oxford, UK: Berg.
   2005. *The Nature of Magic: An Anthropology of Consciousness*. Oxford, UK: Berg.
Gustavsson, Madeleine, and Mark Riley.
   2019. "(R)evolving Masculinities in Times of Change amongst Small-Scale Fishers in North Wales." *Gender, Place & Culture: A Journal of Feminist Geography*. https://doi.org/10.1080/0966369X.2019.1609914.

Haenn, Nora, and Richard R. Wilk.
  2006. *The Environment in Anthropology: A Reader in Ecology, Culture, and Sustainable Living*. New York: New York University Press.
Harper, Sarah, Dirk Zeller, Melissa Hauzer, Daniel Pauly, and Ussif Rashid Sumaila.
  2013. "Women and Fisheries: Contribution to Food Security and Local Economies." *Marine Policy* 39: 56–63. https://doi.org/10.1016/j.marpol.2012.10.018.
Harvey, David, and Jim Perry, eds.
  2015. *The Future of Heritage as Climates Change: Loss, Adaptation, and Creativity*. London: Routledge.
Hastrup, Kirsten, and Frida Hastrup, eds.
  2016. *Waterworlds: Anthropology in Fluid Environments*. New York: Berghahn.
Hemingway, Ernest.
  1952. *The Old Man and the Sea*. New York: Charles Scribner's Sons
Holling, C. S., ed.
  1978. *Adaptive Environmental Assessment and Management*. Toronto: John Wiley & Sons.
Horton, Tom.
  1996. *An Island Out of Time: A Memoir of Smith Island in Chesapeake Bay*. New York: W. W. Norton.
Ingold, Tim.
  2013. *Making: Anthropology, Archeology, Art and Architecture*. London: Routledge.
  2011. *Being Alive: Essays on Movement, Knowledge and Description*. London: Routledge.
  2007. *Lines: A Brief History*. Oxford, UK: Routledge.
  2000. *The Perception of the Environment: Essays on Livelihood, Dwelling and Skill*. London: Routledge.
  1993. "Globes and Spheres: The Topology of Environment." In *Environmentalism: The View from Anthropology*, ed. Kay Milton. Pp. 31–42. London: Routledge.
  1987. *The Appropriation of Nature: Essays on Human Ecology and Social Relations*. Iowa City: University of Iowa Press.
Jackson, Michael, and Albert Piette, eds.
  2015. *What Is Existential Anthropology?* New York: Berghahn.
Johnson, Paula J.
  1992. *The Workboats of Smith Island*. Baltimore: Johns Hopkins University Press.
Khan, Naveeda.
  2014. "The Death of Nature in the Era of Global Warming." In *Wording the World: Veena Das and Her Interlocutors*, ed. Roma Chatterji. Pp. 288–299. New York: Fordham University Press.
King, Sarah.
  2014. *Fishing in Contested Waters: Place and Community in Burnt Church / Esgenoôpetitj*. Toronto: University of Toronto Press.
Kitching, Frances, and Susan Stiles Dowell.
  1981. *Mrs. Kitching's Smith Island Cookbook*. Centreville, MD: Tidewater.

Kohn, Tamara.
   2006. "Conceptualizing Island-Ness." In *Managing Island Life: Social, Economic and Political Dimensions of Formality and Informality in 'Island' Communities*, ed. Jonathan Skinner and Mils Hills. Pp. 79–95. Dundee, Scotland: University of Abertay Press.
   2002a. "Becoming an Islander through Action in the Scottish Hebrides." *Journal of the Royal Anthropological Institute* 8 (1): 143–158.
   2002b. "Imagining Islands." In *World Islands in Prehistory: International Insular Investigations; Fifth Deia International Conference of Prehistory*, ed. William H. Waldren and J. A. Ensenyat. BAR International Series 1095. Pp. 39–43. Oxford, UK: Archeopress.
   1997. "Island Involvement and the Evolving Tourist." In *Tourists and Tourism: Identifying with People and Places*, ed. Simone Abram, Jacqueline Waldren, and Donald V. L. Macleod. Pp. 13–28. Oxford, UK: Berg.
Kopelent Rehak, Jana.
   2019a. "Aging in Place: Changing Socio-Ecology and the Power of Kinship on Smith Island." *Journal of Anthropology and Aging* 40 (1): 43–57. https://doi.org/10.5195/aa.2019.181.
   2019b. "We Live in the Water." *Practicing Anthropology: Journal of the Society for Applied Anthropology* 41 (3): 48–52. https://doi.org/10.17730/0888-4552.41.3.48.
Kopnina, Helen, and Eleanor Shoreman-Ouiment.
   2013. *Environmental Anthropology: Future Trends*. London: Routledge.
Krupnik, Igor, Rachel Mason, and Tonia Horton, eds.
   2004. *Northern Ethnographic Landscapes*. Washington, DC: Smithsonian Institution.
Lambek, Michael.
   1981. *Human Spirits: A Cultural Account of Trance in Mayotte*. Cambridge: Cambridge University Press.
Lansing, J. Stephen.
   2006. *Perfect Order: Recognizing Complexity in Bali*. Princeton, NJ: Princeton University Press.
Lawson, Glenn.
   1988. *The Last Waterman: A True Story*. Alexandria, VA: Washington Book Distributors.
Lee, Kai N.
   1993. *Compass and Gyroscope: Integrating Science and Politics for the Environment*. Washington, DC: Island Press.
Limón, José.
   1994. *Dancing with the Devil: Society and Cultural Poetics in Mexican-American South Texas*. Madison: University of Wisconsin Press.
Lynch, Caitrin.
   2012. *Retirement on the Line: Age, Work, and an American Factory*. Ithaca, NY: Cornell University Press.
Lynch, Caitrin, and Jason Danely.
   2013. *Transitions and Transformations: Cultural Perspective on Aging and the Life Course*. New York: Berghahn.

Mageo, Jeannette.
    2021. "Defining New Directions in the Anthropology of Dreaming." In *New Directions in the Anthropology of Dreaming*, ed. Jeannette Mageo and Robin E. Sheriff. Chapter 1. London: Routledge.

Mageo, Jeannette, and Robin E. Sheriff, eds.
    2021. *New Directions in the Anthropology of Dreaming*. London: Routledge.

Marino, Elizabeth.
    2015. *Fierce Climate, Sacred Ground: An Ethnography of Climate Change in Shishmaref, Alaska*. Fairbanks: University of Alaska Press.

Martine, Lindsey.
    2013. "Narrating Pain and Seeking Continuity: A Life Course Approach to Chronic Pain Management." In *Transitions and Transformations: Cultural Perspective on Aging and the Life Course,* ed. Caitrin Lynch and Jason Danely. Pp. 37–48. New York: Berghahn.

Maurstad, Anita.
    2004. "Cultural Seascapes: Preserving Local Fishermen's Knowledge in Northern Norway." In *Northern Ethnographic Landscapes*, ed. Igor Krupnik, Rachel Mason, and Tonia Horton. Pp. 277–297. Washington, DC: Smithsonian Institution.

Mauss, Marcel.
    1950. *The Gift: Forms and Functions of Exchange in Archaic Societies*. New York: W. W. Norton.

Miller Hesed, Christine D., Elizabeth R. Van Dolah, and Michael Paolisso.
    2020. "Engaging Faith-Based Communities for Rural Coastal Resilience: Lessons from Collaborative Learning on the Chesapeake Bay." *Climatic Change: An Interdisciplinary, International Journal Devoted to the Description, Causes and Implications of Climatic Change* 159: 37–57.

Milton, Kay.
    1993. *Environmentalism: The View from Anthropology*. London: Routledge.

Myerhoff, Barbara G.
    2007. *Stories as Equipment for Living: Last Talks and Tales of Barbara Myerhoff*, ed. Marc Kaminsky and Deena Metzger. Ann Arbor: University of Michigan Press.
    1978. *Number Our Days*. New York: E. P. Dutton.

Norgaard, Kari Marie.
    2011. *Living in Denial: Climate Change Emotions and Everyday Life*. Cambridge, MA: MIT Press.

Orlove, Benjamin S., Ellen Wiegand, and Brian H. Luckman.
    2008. *Darkening Peaks, Glacier Retreat, Science, and Society*. Berkeley: University of California Press.

Paolisso, Michael.
    2003. "Chesapeake Bay Watermen, Weather, and Blue Crabs: Cultural Models and Fishery Policies." In *Weather, Climate, and Culture*, ed. Sarah Strauss and Benjamin S. Orlove. Pp. 61–82. New York: Berg.

2006. *Chesapeake Environmentalism: Rethinking Culture to Strengthen Restoration and Resource Management*. Chesapeake Perspectives Monographs. College Park: Maryland Sea Grant College.

Paolisso, Michael, Christina Prell, Katherine J. Johnson, Brian Needelman, Ibraheem M. P. Kahn, and Klaus Hubacek.
2019. "Enhancing Socio-Ecological Resilience in Coastal Regions through Collaborative Science, Knowledge Exchange and Social Networks: A Case Study of the Deal Island Peninsula, USA." *Socio-Ecological Practice Research* 1 (2): 109–123.

Petryna, Adriana.
2022. *Horizon Work: At the Edges of Knowledge in an Age of Runaway Climate Change*. Princeton, NJ: Princeton University Press,

Pink, Sara.
2009. *Doing Sensory Ethnography*. Los Angeles: Sage.

Ponkrat, Bob, and Laura Stocker.
2011. "Anthropology, Climate Change and Coastal Planning." In *Environmental Anthropology Today*, ed. Helen Kopnina and Eleanor Shoreman-Ouiment. Pp. 179–194. London: Routledge.

Power, Nicole G.
2008. "Occupational Risks, Safety and Masculinity: Newfoundland Fish Harvesters' Experiences and Understandings of Fishery Risks." *Health, Risk & Society* 10 (6): 565–583. https://doi.org/10.1080/13698570802167405.

Probyn, Elspeth.
2016. *Eating the Ocean*. Durham, NC: Duke University Press.
2014. "Women Following Fish in a More-than-Human World." *Gender, Place & Culture: A Journal of Feminist Geography* 21 (5): 589–603. https://doi.org/10.1080/0966369X.2013.810597.

Pugh, Jonathan, and David Chandler.
2021. *Anthropocene Islands: Entangled Worlds*. London: University of Westminster Press.

Ram, Kalpana, and Christopher Houston.
2015. *Phenomenology in Anthropology: A Sense of Perspective*. Bloomington: Indiana University Press.

Rappaport, Roy A.
1979. *Ecology, Meaning and Religion*. Richmond, CA: North Atlantic Books.

Roscoe, Paul.
2014. "A Changing Climate for Anthropological and Archeological Research? Improving the Climate-Change Models." *American Anthropologist* 116 (3): 535–548.

Saul, Shura.
1974. *Aging: An Album of People Growing Old*. New York: John Wiley & Sons.

Scarry, Elaine.
1999. *On Beauty and Being Just*. Princeton, NJ: Princeton University Press.
1985. *The Body in Pain: The Making and Unmaking of the World*. New York: Oxford University Press.

Schaie, K. Warner, and Carmi Schooler, eds.
  1998. *Impact of Work on Older Adults*. New York: Springer.

Schilling, Natalie.
  2017. "Smith Island English: Past, Present, and Future—and What Does It Tell Us about the Regional, Temporal, and Social Patterning of Language Variation and Change?" *American Speech: A Quarterly of Linguistic Usage* 92 (2): 176–203. https://doi.org/10.1215/00031283-4202020.

Shewery, Teresa.
  2015. *Hope at Sea: Possible Ecologies in Oceanic Literature*. Minneapolis: University of Minnesota Press.

Shultz, Kenneth S., and Gary A. Adams, eds.
  Aging and Work in the 21st Century. New York: Routledge.

Sokolovsky, Jay.
  1990. *The Cultural Context of Aging: Worldwide Perspectives*. New York: Bergin & Garvey.

Sponsel, Leslie E.
  2012. *Spiritual Ecology*. Santa Barbara, CA: Praeger.
  2011. "The Religion and Environment Interface: Spiritual Ecology in Ecological Anthropology." In *Environmental Anthropology Today*, ed. Helen Kopnina and Eleanor Shoreman-Ouiment. Pp. 37–55. London: Routledge.

Steward, Julian Haynes.
  1955. *Theory of Culture Change: The Methodology of Multilinear Evolution*. Urbana: University of Illinois Press.

Strang, Veronica.
  2015. *Water: Nature and Culture*. London: Reaktion Books.
  2004. *The Meaning of Water*. Oxford, UK: Berg.
  1997. *Uncommon Ground: Cultural Landscape and Environmental Values*. Oxford, UK: Berg.

Strauss, Sarah, and Benjamin S. Orlove.
  2003. *Weather, Climate, and Culture*. Oxford, UK: Berg.

Stoller, Paul.
  1997. *Sensuous Scholarship*. Philadelphia: University of Pennsylvania Press.
  1989. *The Taste of Ethnographic Things: The Senses in Anthropology*. Philadelphia: University of Pennsylvania Pres.

Tsing, Anna Lowenhaupt.
  2015. *The Mushroom at the End of the World: On the Possibility of Life in Capitalist Ruins*. Princeton, NJ: Princeton University Press.

Turner, Victor Witter.
  1992. *The Anthropology of Performance*. New York: PAJ.
  1977. *The Ritual Process: Structure and Anti-Structure*. Ithaca, NY: Cornell University Press
  1974. *Dramas, Fields, and Metaphors: Symbolic Action in Human Society*. Ithaca, NY: Cornell University Press.
  1969. *The Ritual Process: Structure and Anti-Structure*. Ithaca, NY: Cornell University Press.

van Gennep, Arnold.
> 1961. *The Rites of Passage*. Chicago: University of Chicago Press.

Van Heekeren, Deborah.
> 2012. *The Shark Warrior of Alewai: A Phenomenology of Melanesian Identity*. Wantage, UK: Sean Kingston.

Waldren, William H., and J. A. Ensenyat, eds..
> 2002. *World Islands in Prehistory: International Insular Investigations; Fifth Deia International Conference of Prehistory*. BAR International Series 1095. Oxford, UK: Archeopress.

Warner, William.
> 1976. *Beautiful Swimmers: Watermen, Crabs, and the Chesapeake Bay*. Atlantic Monthly Press Book. Boston: Little, Brown

White, Christopher.
> 2009. *Skipjack: The Story of America's Last Sailing Oystermen*. New York: St. Martin's.

Index

agency, exercised by older islanders, 74–75
aging, xii, xiv, xv, 14, 16, 24, 61, 76–77, 102–103, 110, 191
anthropology, 10–14, 17–24, 51–55, 153, 163, 192
art, 175; Surrealism, 176

Bateson, Gregory, 36, 46, 163, 174
beauty, in landscape, 61, 108
Berkes, Fikret, 6, 10, 50, 163
Bible, 101
birds, 90, 176–177
blue space (places around water), 11, 35–36
boats, working, 60, 70–71
body, health of, 117–118, 121

Chesapeake Bay, vii–ix, x–xii
childhood, 57–77, 80–82, 103
Christianity, 44–46, 102–103
citizenship, 3
climate change, xvii, xiii, xvii, 1–4, 6–12, 7–13, 48, 53
comedy, 105
community, 131–132, 147, 150, 168–170, 179, 182, 188
cosmological knowledge, xiii, xvi, 5–6
crabbing, 95, 120, 123
crabs, 48, 50, 58, 68, 70, 157
Crate, Susan, 9–10
creativity, xiv, xv, 149, 161, 164, 175–174, 183–184, 188, 190

death, views on, 5, 48, 102, 104–116

ecological change, 2–5, 9–10, 54, 87–89, 179
ecological knowledge, xiii, xvi, 5, 6, 11–12, 15, 39–42, 46, 53–54
economy, xiv
emotions, 6, 36–37, 39, 61–62
environmental knowledge, 4–5, 6, 32, 35, 41, 54–55, 63–73, 145–155
erosion, of coastline, xiv, 84, 114, 78–95
ethics, 91, 146
ethnographic fieldwork, 16–24, 191–192
ethnography, xv, 8–9, 17, 22–23
ethno-poetry, 109, 111, 112–113, 194–198
experiential learning, 58, 60, 64, 66, 68–69

faith, xiv, 6, 44–46, 58, 87, 102
family photo albums, 16–23
fieldwork, anthropological, 16–24, 191–192
flooding, xvii, 7, 78–94, 173
food, 57–77, 159–164
freedom, personal, 64, 72
funeral rites, 105–107
future of Smith Island, 7

gardening, 85–88
gender, 63–75, 140–164, 168
gift-giving, 115–116, 151, 193

happiness, 62, 165, 182
health, 117–139, 179, 180
heritage, 4–5, 33, 58–59, 164

hope, 7, 11, 15
humor, 140–146, 151–152
Hurricane Sandy, 1–2, 7

imagination, 35, 38–39, 154
Ingold, Tim, 11–12, 39, 45–46, 53, 163, 174

jokes, 143–145, 157
joy, 62, 150, 165

kinship, 44, 57–67, 70, 74, 96–109, 112–115, 136–139
knowledge: cosmological, xiii, xvi, 5–6; ecological, xiii, xvi, 5–6, 58, 72; environmental, 4–5, 6, 32; gendered, 75–77, 154–164; humorous, 140–145; sensory, xiii, xvi, 5–6, 18–23, 57–72, 159–160, 169, 190

land: changing, 12, 59, 78–95, 114–115; management of, 78, 84, 94
language, 42; and play, 143–145, 105
life course, 65, 73, 96–116, 118, 124, 127–139, 166–169

magic, 20, 46, 139, 166, 169, 173–75, 177–178, 180, 184
media, 21–22
medical care: history of, on Smith Island, 125–127

nursing home, living in, 99–100, 142

oral traditions, xiv, xv, 35, 47–51, 143
oyster farming, 6
oysters, xii, 120, 123

pain, 119, 133–137
Paolisso, Michael, 6, 41, 45
parody, 144–146, 149
phenomenology, 7, 11–12, 14–15, 114

photography, 18–23, 57–58, 78–81, 83, 99, 110, 137–138, 157–158, 141
placemaking, x, xii, 6, 25, 32
play, as social communication, 68–70
poetics, xiv, xv, 16, 190–199
poetry, by islanders, 16–17, 110–113
political discourse, 2–5

resilience, 2, 192
rites of passage, 43, 59, 63–70
ritualized laughter, 146
rituals, 146, 153, 163

seascape, 59, 62, 66–67
sense of self, 18, 35, 53, 57–77, 140, 150–151, 152–153
sensory knowledge, xiii, xvi, 5–6, 18–23, 25, 57–72, 159, 169
socioecology, 7, 9–13
soundscape, 90
spirits, 177
storytelling, 16, 25, 35, 47–51, 143, 145
Strang, Veronica, 36–37
subjectivity, 80
sustainability, 6, 85–87, 117–123

taste, of food, 87–88, 90–92
telemedicine, 127–128, 130–131

uncertainty, of life conditions, 4, 36, 45, 48

vulnerability, 2–3, 117–139

water, 11, 35–36, 47–49
watermen, 3, 5–6, 51–52, 58–59, 122, 124
weather, xii, xvii, 1–2, 3–5, 14, 32, 35–36, 39–43, 52–53, 113, 116, 169; knowledge of, 5, 35–36, 41–42
wind, 37–40
work, on Smith Island, 57–58, 67, 118, 121, 141–142, 150, 156–158, 167–170